The Unforeseen Forces: Unveiling Hidden Influences on Solidification

Shivani

Table of Contents

1   Introduction   1

2   Background   4

   2.1   Overview   4

   2.2   Rapid Solidification Studies   4

   2.3   Pyrometry Theory   8

   2.4   High-Speed Imaging   12

   2.5   Geometry-Based Solidification Model   13

   2.6   Retained Damage Model   19

   2.7   Thermophysical Properties   23

3   Methodology   26

   3.1   Overview   26

   3.2   Experimental Setup   27

   3.3   Sample Preparation   30

   3.4   Data Collection Procedure   32

   3.5   Computing Delay Time   35

4   Modeling   39

   4.1   Overview   39

   4.2   Surrogate Modeling (MHD)   39

   4.3   Modeling the Influence of Convection by a Saturation Approach   49

   4.4   Model Validation by Dimensionless Predictions   52

5   Discussion   60

   5.1   Overview   60

5.2	Convection Model Comparison	60
5.3	Alloy Behavior Comparison	62
5.4	Comments on Model Validity	64
6	Conclusions and Future Work	66
6.1	Utility of MHD Modeling Approach	66
6.2	Damage Model Comparison	66
6.3	Alloy Comparison	67
6.4	Model Validation	68
6.5	Future Work	69
7	References	71
	Appendices	75

Chapter 1. Introduction

In an industrial setting, operations such as casting and additive manufacturing rely on an accurate understanding of a wide array of materials and their varied reactions to the thermal environment in which they are processed. While plastics have retained a firm foothold in consumer industries for their adaptability to nearly any form and their light weight, there are numerous applications where low strength polymers simply cannot withstand the large stresses or strains that their product will be subjected to. This is especially true in the aerospace industry where tools and components must be equally as strong as they are formable, light, and inexpensive. For these purposes, there is no better material than steels or iron based alloys. While many of the key thermophysical properties for these materials have been well reported in literature [1], fine control of their reactions on a molecular level at specific stages during the manufacturing process is typically gleaned from years of experimentation and is often a closely guarded trade secret of the select few manufacturing firms responsible for their production.

In steel and iron alloys, material composition and structure must be carefully controlled during solidification to ensure that the final product meets specific requirements. In bulk, a material might exhibit no signs of failure, but on the atomic scale, a collection of minor defects could significantly compromise the strength or durability of the final product. It is well known that these properties may be easily influenced by externally applied forces, fields, or thermal loads. External factors such as gravitational forces can occur whether the manufacturer intends for them to be present or not and can have a significant impact on the strength and quality of a resulting product, tool, or component. The specifics of interactions between applied and unintentional influences on the solidification process is currently a matter of debate which is the driving question intended to be answered by this book. It is important that scientists better understand controllable effects of solidification in an effort to make publicly available information that is capable of greatly improving the strength of manufactured products for both consumer and aerospace purposes.

In pursuit of this endeavor, emphasis must be placed on the specifics of phase selection, growth, and microstructural evolution throughout the course of solidification. It has been well documented in the literature that the duration of a phase transformation is heavily reliant on the quantity of total available Gibbs free energy contained to the system [2-12]. One of the largest components of this free energy can be tied directly to the amount of internal convective stirring within the material at a given time. If precision control of convective stirring can be achieved, it becomes possible to influence the duration of phase growth, effectively altering the microstructure and resulting materials properties. Unfortunately, while stirring can be applied intentionally in ferrous alloys through the application of a magnetic field, it can also be driven by uncontrollable gravitational effects or flows produced by kinetic motion around the surfaces of a mold.

Space-based research offers investigators the chance to isolate and observe convective influence on a solidifying sample in the absence of gravitational or external forces thanks to a zero-gravity levitation environment. Therein, the manual adjustment of input control voltages to an electromagnetic levitation (EML) coil can apply a quantifiable amount of stirring. The effects of which are directly comparable to ground based levitation facilities where the only effects on a sample are due to gravitational effects. The motivation for this book is to reinforce the connection between convective forces within a sample and its effects on phase growth by first linking ISS-EML sample behavior to the Retained Damage Model (RDM) [13] and then developing optimized parameters for two convection damage models: the Read-Shockley Dislocation Energy and the Damage Saturation Model approaches. These models will be compared against collected experimental data and adjusted to highlight key changes between different ferrous alloy compositions. The final data set considered in this analysis spans a wide range of both convective conditions and undercoolings, lending itself to a great many applications.

Chapter 2. Background

2.1 Overview

The information detailed in the following section will provide a literature review around the key concepts critical to the microstructural evolution of ferrous alloys. Specifically, the equipment used by researchers to capture these events will be covered, along with a presentation of common thermophysical properties of common manufacturing alloys. Two models that attempt to quantify the driving factors behind material transformations during solidification will be discussed, and their grounding equations derived. The first method, the Classical Nucleation Theory (CNT) applies growth kinetics approximated by the Lipton-Kurz-Triveldi/ Boettinger-Coriell-Triveldi (LKT/BCT) Model to develop a geometry-based prediction of solidification rates, while the second method, the Retained Damage Model (RDM) follows a thermodynamic approach to investigate the application of controllable parameters such as stirring and its impact on transformation delay times. The RDM method alone is able to leverage predictions as a tool for enhancing microstructural development and material properties in everyday products by selectively adding or removing stirring to achieve a desired result.

2.2 Rapid Solidification Studies

The transformation from solid to liquid in ferrous alloys is notably reversible, with the steel reverting to a solid state when enough thermal energy is extracted from the environment to lower internal temperature below the liquidus. The re-solidification is once again gradual, with crystals nucleating in small pockets of low temperature alloy and then growing outward into the surrounding liquid by picking up loose atoms and adding them to a branching dendritic structure. Typically, the initial nucleation will occur around a defect or oxide. There are several unique crystal structures which can nucleate independently of one another. Among the most common are face centered cubic (FCC), base centered cubic (BCC), and hexagonal close packed (HCP) pictured in Figure 1 [1, 14].

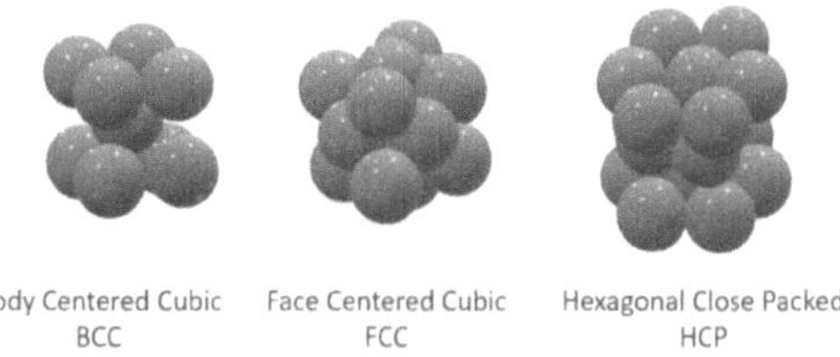

Figure 1 – Common Crystal Structures

During solidification within the liquid there is phase competition between these structures for control over the final material composition. The internal energy requirement for a given phase structure to form, along with its ease of retaining atoms, and accepting new atoms at a given temperature each play a role in which phase will overtake the others. The phase that is ultimately selected is dependent on the material composition [13, 14]. When the primary phase growth occurs, it will often appear as a pyramidal structure with its peak growing along preferred crystallographic directions dictated by anisotropic surface energies along the boundary between the metal and its external environment. The intersection of this geometry with a curved surface will yield a diamond-like shape that will grow larger as the temperature is lowered. The growth rate of the preferred directions appear as corners of a polygon as this phase grows across the surface. This growth is known to be a function of the temperature at which the sample was cooled to at the instant of nucleation relative to the liquidus temperature [14]. This temperature difference is known as undercooling and is an uncontrollable aspect of solidification. Eventually the primary phase growth morphology will proceed to envelop the entire metal surface. The duration from when nucleation initiates to when the metal is consumed by a given phase is characterized by a rise in temperature that is known as recalescence. During liquid recalescence the temperature rise results from conversion of atomic translational kinetic energy in the liquid to vibrational energy in the solid crystal lattice. During solid recalescence the rise is generated by a release of free energy stored in the metastable solid as atoms rearrange to their stable phase crystal structure. Additional energy contributions may come from free energy stored in defects within the crystal structure that are put under pressure and locked in place during cooling of the liquid region as it solidifies [14]. In ferrous alloys, whichever phase occurs first is known as the 'metastable' phase because the recalescence temperature rise partially remelts the metal allowing for further secondary phase transformations to take root.

Once the temperature has sufficiently risen, a secondary 'stable' phase quickly nucleates and emerges from within the metastable phase. It will begin to grow outward into the pre-existing metastable mushy zone with a growth independent of the primary phase growth rate. The secondary recalescence process initiated by this growth will release less internal energy than the phase preceding it, resulting in a much smaller temperature rise than that of the primary phase. The crystal structure of the secondary phase is often dense, with fine grains. The temperature at the end of the formation will be nearly at the liquidus, beyond which no further heat will be generated, and cooling will resume until the entire metal has reached a thermal equilibrium with the surrounding environment, effectively locking the final crystal structure in place and characterizing bulk material properties such as strength and toughness. In the immediate aftermath of secondary phase nucleation, there will be a brief period where the primary phase has not yet completed its growth into the surrounding liquid. During this time, the growth rates of the primary phase growing into the liquid region and secondary phase growing into the metastable mushy zone of the liquified primary phase are in direct competition during a process known as phase selection. If the secondary stable phase can grow faster than the metastable phase it will overtake it and break-out into the undercooled liquid to form a very different microstructure [15]. The timing between primary and secondary phase nucleation is known as a "delay time". It is influenced by the available free energy within the sample at the point of nucleation which is in turn a function of undercooled temperature relative to the liquidus at the time of the primary phase nucleation. It has been found that any additional energy added to the metal by external means can also affect the timing of nucleation events.

According to the Retained Damage Model, deeper undercoolings invariably lead to more available free energy being stored within defects in the lattice as pressure mounts on these singular regions [1, 8]. With more energy available to drive nucleation and subsequent growth, growth rates for both primary and secondary phases are accelerated, and secondary nucleation can occur much sooner than in a shallow undercooled metal. The result is a smaller delay time, which will yield a finer grain structure thereby influencing bulk material properties and the overall quality of a product's surface. Undercooling is an uncontrollable aspect of solidification, so it provides little benefit when attempting to actively influence microstructure. However, the process of solidification can be controlled by adding an externally applied energy source. One example of this would be stirring the molten liquid during solidification which imparts kinetic energy directly to a metal through flows resulting from internal convection. The more forceful the stirring, the more energy that will be imparted, leading to significantly faster growth of a stable phase and hence shorter delay times. The effect of convection on delay time overshadows the influence of both primary and secondary phase formation energy, making it an ideal means to directly influence microstructure.

2.3   Pyrometry Theory

The cooling of a material during solidification can be recorded using high frequency pyrometry measurements which are capable of characterizing surface temperature at any given point in time. It is known that temperatures vary throughout a material and across the surface due to unequal application of heating. One previous study which modeled temperature gradients through a finite element analysis has shown that the surface of a spherical FeCrNi sample during solidification can expect a variation of 0.1 degrees when the sample is not actively heated during solidification, and 3.2 degrees when the sample is held at a constant temperature of 1500 Kelvin during the simulation to represent worst case heating during solidification [16]. It was concluded that in both cases the temperature distribution across the surface was negligible and that it would not influence phase growth or selection during solidification [16].

A typical pyrometer will sample measurements at a rate of 100 Hz [17]. Such is the case for the International Space Station's Axial Camera and Pyrometer mounted to the electromagnetic levitator Columbus module [17]. Sampling at high frequency allows for clear resolution of key temperatures during recalescence events. At 100 Hz the undercooled temperature at the instant of primary phase nucleation can be clearly seen. In some steels such as the ones that will be discussed in this work, the timing between primary and secondary phase transformations during solidification is so short and temperature differences so insignificant that pyrometry is unable to distinguish between them. Thus, in these cases only the final temperature of the material at the end of the secondary phase transformation will be recorded, displaying as a single peak at the end of a linear rise from the undercooled temperature over the span of two temperature readings. In other materials where delay is longer than the pyrometry refresh rate, temperature differences will be more pronounced between the two phases. In these cases, the end of the first phase will present as a peak, followed by a slight dip as the sample begins to cool, then a second peak at a higher temperature for the second phase.

The output of a pyrometer is a raw temperature measure that is typically recorded in degrees Celsius. Each measurement is associated with a time stamp when it was taken which allows for easy calculation of cumulative time. Temperature can be plotted against time to better visualize changes over the course of any solidification experiment. This requires the time stamp output of a pyrometer to be converted into a measure of time. To obtain a time reading in seconds, the time stamp can be multiplied by a series of conversion factors starting with 24 hours/day, then 60 minutes/hour, and finally 60 seconds/minute. The result will display seconds in the day up until that measurement was taken. Subtracting the time in seconds of the first measurement from all subsequent measurement time stamps will then yield the time of a measurement relative to the start of any experimentation. For cases where the time when a solidification was triggered is known, temperatures relative to that event will be easier to locate.

One source of error in temperature readings from pyrometry is known as an emissivity shift. These increases in emitted radiative energy closely follow rises in temperature, and lead to inflated temperature readings at specific points along a thermal profile, they can disrupt the accuracy of critical temperature measures and thereby impact experimental results. Pyrometry results affected by emissivity shifts or similar errors can be easily corrected by comparing the recorded temperature at the instant of a known phase transformation to the value reported in literature for the specific material composition being studied. The liquidus temperature lends itself well to this approach as it is commonly reported for a wide range of material compositions. The point at which a material reaches the liquidus is also visible on a temperature – time plot by a transition from a shallow temperature rise to a steeper one when the sample has become fully liquid under constant heating. The entire temperature plot can be compressed by the difference between the observed and reported literature value of this point. Temperature correction relies on an assumed emissivity ε_λ relative to the surrounding atmosphere which is required to compute the black body temperature T_b through implementation of Equation 1 for a gray body. Here, C_2 is Planck's second radiation constant, λ is the working wavelength of the pyrometer, and T is the true temperature [18]. It should be noted that emissivity is inversely proportional to the blackbody temperature, so a hotter region on a solidifying metal with higher emissivity will yield a similarly lower blackbody temperature.

$$\frac{1}{T_b} = \left(\frac{1}{T} - \frac{\lambda}{C_2}\right) \ln \varepsilon_\lambda \tag{1}$$

When $\lambda\, T_b$ is much smaller than C_2, the materials radiance I can be computed from Wein's approximation to Planck's radiation equation as shown in Equation 2 [19]. Here k is a constant parameter, and C_1, is Planck's first radiation constant.

$$I = k\frac{C_1}{\lambda^5} \frac{1}{\exp\left[\frac{C_2}{\lambda T_b}\right]-1} = k\frac{C_1}{\lambda^5} \exp\left[-\frac{C_2}{\lambda T_b}\right] \tag{2}$$

In the event that the pyrometer is not providing a true reading of temperature, the observed temperature at a key point such as the liquidus T_{PL} will differ from the known liquidus temperature T_L for that alloy taken from literature. Using the liquidus temperature in Equation 1 and Equation 2 will yield a reference radiance I_L which differs from the radiance I_{PL} calculated instead from the observed liquidus temperature. Similarly, at any observed temperature above or below the liquidus, the observed temperature T_P will produce a radiance I_P which will differ from the true radiance I. Forming a ratio between any two radiances mentioned above and taking the natural log of both sides as has been done in Equation 3 for the true radiance and the true radiance at the liquidus using the temperature from literature. The relationship between their respective temperatures is dependent on emissivity associated with each of the temperatures used [19].

$$\ln\left(\frac{I}{I_L}\right) = -\frac{C_2}{\lambda}\left(\frac{1}{T} - \frac{1}{T_L}\right) + \ln\left(\frac{\varepsilon_\lambda}{\varepsilon_{\lambda L}}\right) \tag{3}$$

If two radiances are compared using temperatures which have been obtained by the same measurement method at the same location, then their emissivity's will be equivalent and the final term on the right-hand side of Equation 3 can be eliminated. Such is the case with the ratio between I, I_L and for the ratio between I_P, I_{PL}. These two ratios between a given temperature and a liquidus temperature can be set equal, yielding the following expression Equation 4. The constants C_2 and λ are equivalent on both sides, leaving a relationship between four temperatures, all of which are known except for the true temperature T representing the correction of T_P such that at the liquidus T_{PL} equals the true liquidus from literature T_L [19].

$$-\frac{C_2}{\lambda}\left(\frac{1}{T} - \frac{1}{T_L}\right) = -\frac{C_2}{\lambda}\left(\frac{1}{T_P} - \frac{1}{T_{PL}}\right) \tag{4}$$

Rearranging to solve for that unknown true temperature T yields the temperature correction Equation 5 [18].

$$\frac{1}{T} = \left(\frac{1}{T_P}\right) + \left(\frac{1}{T_L} - \frac{1}{T_{PL}}\right) \tag{5}$$

This equation can be applied to every temperature reading T_P in a pyrometry feed where the liquidus point T_{PL} can be identified on a plot of uncorrected temperature vs. time. For this point to be found, there must be a relatively clean pyrometry feed with little noise that would otherwise obfuscate this critical temperature. Plotting the calculated true temperature against time will provide more accurate measures of key temperatures and undercooling temperature differences required for solidification analysis.

2.4   High-Speed Imaging

In addition to measurements taken with pyrometry, solidification can also be captured using high-speed imagery. At high frame rates, changes in brightness of a material around nucleation events can be associated with temperature rises driven by phase growth. This allows for the stages of primary and secondary phase growth to be directly observed. Velocities can be recorded by tracking points of interest as they translate between frames. To capture something as fast as a secondary phase transformation in steel alloys, frame rates around 30,000 or 50,000 FPS are required depending on the alloy [1].

The necessity for the high-speed camera to capture phase transformation at the instant of nucleation when the phase is merely a small cluster of atoms requires image detail sufficient to differentiate phase growth of at least one pixel per frame. This requires a frame resolution of at least 250^2 pixels. The more pixels available, the closer to nucleation a phase can be recognized and the more accurate its initial growth velocity can be recorded. The high-speed camera must also be placed sufficiently close to the sample subject. In the case of the ISS-EML facility, the radial camera is mounted at a distance of about 256mm between the camera lens and the sample's center, while a secondary axial camera is mounted at an approximate distance of 266mm to 269mm. The radial camera operates at a higher frame rate and is thus best suited for the capture of high-speed video for solidification experiments.

As accurate as pyrometers can be when properly calibrated, they are still subject to sources of error. One potential impact to the accuracy of pyrometry readings is transmissivity of the optical path where evaporated particles from the material come between the sensor and its target. Just as sample evaporation can influence the accuracy of pyrometry readings, it can also impact the clarity with which a camera can capture surface events. This must be considered in the design of a facility used to conduct solidification experiments.

2.5   Geometry-Based Solidification Model

Solidification studies directed at the identification of microstructural evolution through use of pyrometry and high-speed imagery are scarce in the literature with only a handful of theories being put forward as to the true influences on phase selection, growth, and secondary transformation timing. One study has suggested that timing of phase transformations can be inferred by first considering the Lipton-Kurz-Triveldi (LKT) Model which characterizes the geometry of dendritic structures and approximates their microscopic growth kinetics through the identification of atom attachment rates which can in turn be modeled through a comparison with experimental data of dendrite tip formation rates [20,21]. Once geometric and kinetic approximations are made, the Classical Nucleation Theory (CNT) then allows for delay times to be produced as a function of geometric parameters [22]. This model provides a theoretical approach to understanding the development of secondary phase transformations and a means to calculate the exact timing of secondary phase transformations when solidification temperature is known [22]. The CNT model also accounts for non-equilibrium effects and can be further modified to include dendrite growth kinetics using the Boettinger-Coriell-Triveldi (BCT) Model [21]. One key limitation of the CNT model is an absence of controllable parameters such as the influences of convection which would allow for direct influence of behavior.

The LKT/BCT Model for evaluation of the velocity associated with growth of dendrites into undercooled liquids as required for the CNT can be derived starting with the development of Equation 6 for the ratio of enthalpy change relative to temperature change [20]. It is a function of thermophysical properties for heat of fusion of the metastable phase $\Delta H_{f,M}$, the difference in liquidus temperatures between the two phases ΔT_{SM} and the heat capacity of the liquid $C_P{}^L$. Energy absorption into the dendrite tip from the surrounding liquid taken per unit volume q_{abs} is also required as it is a key driver of growth, along with the ratio of volume solid to volume liquid V_R of the fluid into which the dendrite growth occurs. q_{abs} and V_R can each be approximated in Equation 7 as functions of $\Delta H_{f,M}$, ΔT_{SM}, $C_P{}^L$, in addition to the heat of fusion of the stable phase $\Delta H_{f,M}$, the growth velocity V and the parameter J_S representing heat flux away from the dendrite tip perpendicular to dendritic growth. [20].

$$\frac{\partial H}{\partial T} = \frac{q_{abs}}{\Delta T_{SM}} = C_P^L + \frac{\Delta H_{f,M}}{\Delta T_{SM}} \frac{1}{\frac{1}{V_R}+1} \tag{6}$$

$$\text{Where} \quad V_R = \frac{\Delta H_{f,S} - \frac{J_S}{V} - C_P^L \Delta T_{SM}}{\Delta H_{f,M} + C_P^L \Delta T_{SM}} \quad \text{and} \quad q_{abs} = C_P^L \Delta T_{SM} + \Delta H_{f,M}\left(\frac{1}{\frac{1}{V_R}+1}\right) \tag{7}$$

An estimation of undercooling is necessary for the LKT Model. It can be computed as the sum of undercooling components each associated with a unique driving force. The three initial components are due to thermal effects, curvature effects, and solutal effects (ΔT_t, ΔT_r, ΔT_c) respectively. The LKT Model for undercooling can be supplemented with an additional undercooling component influenced by growth kinetics ΔT_k as taken from the BCT Model. The final Sum is given in Equation 8 where each component is a function of dendrite growth velocity which can be solved for at a given undercooling. From dendrite growth velocity, a relationship for delay time can be obtained. [21]

$$\Delta T_T = \Delta T_t + \Delta T_r + \Delta T_c + \Delta T_k \tag{8}$$

The first component ΔT_t can be modeled by Equation 9. It is a function of the Ivantsov function Iv as expanded in Equation 10 to include the thermal Peclet number P_t which is the ratio of dendrite growth in the form of velocity times dendrite tip radius $V*R$ to thermal diffusivity ($2\alpha_L$). [20].

$$\Delta T_t = Iv\,(Pt)\frac{1}{\left(\frac{\partial H}{\partial T}\right)}\left(\Delta H_f - \frac{Js}{V}\right) \tag{9}$$

$$\text{Where}\ \ Iv(P_t) = \left(\frac{VR}{2\alpha_L}\right)e^{\left(\frac{VR}{2\alpha_L}\right)}E_1\left(\frac{VR}{2\alpha_L}\right) \tag{10}$$

The second term considering curvature effects is calculated in Equation 11. It is a function of the Gibbs-Thompson coefficient Γ which is in turn dependent on the quantity of energy at the interface between solid and liquid volumes σ and the thermophysical property, entropy of fusion ΔS_f. [20]. R can be approximated using Equation 12. [22].

$$\Delta T_r = 2\frac{\Gamma}{R}\ \ \text{where}\ \ \Gamma = \frac{\sigma}{\Delta S_f} \tag{11}$$

$$R^2 = \frac{\Gamma/\sigma^*}{(m_L G_C \xi_C)-(\frac{1}{2}G_L \xi_L)} \tag{12}$$

The third term deals with solutal effects and is expanded in Equation 13. Where m_V can be expanded in Equation 14 to be a function of concentration of the solute C_o and the solutal Peclet number which is nearly identical to the thermal Peclet number except for the solutal diffusivity D_o being used instead of thermal diffusivity. Two additional factors reliant on the phase diagram are required to compute solutal undercooling. Checking the slope of the liquidus line provides a measure of m_L, which must be adjusted to compensate for changes in slope m_v due to growth kinetics. m_v can be calculated from the partitioning coefficient k which is in turn a function of the equilibrium partition coefficient Ke, the initial solute atom fraction X_o, and the ratio of solutal diffusivity to atomic spacing $V_d = D_o / a_o$. [20].

$$\Delta T_C = m_L C_o \left[1 - \frac{m_v}{m_L[1-(1-k)\, Iv\left(\frac{VR}{2D_o}\right)]} \right] \tag{13}$$

$$\text{Where } m_v = \left[1 + \frac{\left(ke - k\left(1 - \ln\left(\frac{k}{ke}\right)\right)\right)}{1-ke} \right] \text{ and } k = \frac{ke + \frac{V}{V_d}}{1-(1-Ke)X_o + \frac{V}{V_d}} \tag{14}$$

The final undercooling component dictated by BCT is the growth kinetics term which is expanded in Equation 15 . It is a function of the universal gas constant $\bar{R}$ and dendrite growth velocity V, and the kinetic rate v_o along with previously mentioned thermophysical properties. Two variations of this equation exist; the first assumes that there is some quantity of Js previously defined as heat flux perpendicular to dendrite growth influencing growth kinetics. In this case, J_s must be included in the calculation for ΔT_k. Otherwise, if J_s is assumed to not be present, then the second equation where it is absent can be implemented. [20].

$$\Delta T_k = \frac{V}{\left(\frac{V_o\left(\Delta H_{f,s} - \frac{J_s}{V}\right)}{\bar{R}\, T_m^2}\right)} \quad \text{or} \quad \Delta T_k = \frac{V}{\left(\frac{\Delta H_f\, V_o}{\bar{R}\, T_m^2}\right)} \text{ without } Js \tag{15}$$

With all undercooling terms having now been defined as functions of velocity, the total undercooling Equation 8 can be obtained for any given growth velocity. Velocity must now be obtained using an iterative solution for Equation 16.

$$-\Gamma \omega^2 - [\overline{K_L} G_L \xi_L + \overline{K_S} G_S \xi_S] + m_L G_c \xi_c = 0 \tag{16}$$

Which is a function of the following 3 stability parameters in Equations 17, 18, 19 and includes the velocity dependent Peclet numbers, a stability parameter σ^* , ratio of thermal diffusivity $\bar{\alpha}$ and the weighted solid and liquid conductivities $\bar{k_L}$ and $\bar{k_S}$ shown in Equation 20 [20].

$$\xi_L = \frac{(1+\bar{K})\left(-1+\sqrt{1+\frac{1}{\sigma^* P_T^2}}\right)}{\bar{K}\left(1+\sqrt{1+\frac{1}{\sigma^* P_T^2}}\right)+\left(-\bar{\alpha}+\sqrt{\bar{\alpha}^2+\frac{1}{\sigma^* P_T^2}}\right)} \tag{17}$$

$$\xi_S = \frac{(1+\bar{K})\left(\bar{\alpha}+\sqrt{\bar{\alpha}^2+\frac{1}{\sigma^* P_T^2}}\right)}{\bar{K}\left(1+\sqrt{1+\frac{1}{\sigma^* P_T^2}}\right)+\left(-\bar{\alpha}+\sqrt{\bar{\alpha}^2+\frac{1}{\sigma^* P_T^2}}\right)} \tag{18}$$

$$\xi_c = 1 + \frac{2k}{1-2k-\sqrt{1+\frac{1}{\sigma^* P_C^2}}} \tag{19}$$

$$\text{Where } \quad \bar{k_L} = \frac{k_L}{k_L+k_S} \quad \text{and} \quad \bar{k_S} = \frac{k_S}{k_L+k_S} \tag{20}$$

in addition to the wavenumber ω^2 shown in Equation 21, and thermal concentration gradients GL, GS and GC in Equation 22, Equation 23, and Equation 24 which are also functions of the Peclet numbers and the Ivantsov function [20].

$$\omega^2 = \frac{1}{\sigma^* \lambda^2} \tag{21}$$

$$G_L = \frac{-2P_t}{C_P^L R}\left(\Delta H_f - \frac{J_S}{V}\right) \tag{22}$$

$$G_S = \frac{J_S}{K_S} \tag{23}$$

$$G_S = \frac{-2C_0 P_C(1-k)}{R(1-(1-k)Iv(P_C))} \tag{24}$$

Using these quantities and isolating for V in Equations 15 allows a relationship for phase growth velocity to be approximated by an iterative solution to the quadratic equation shown in Equation 25 [20].

$$AV^2 - BV + C = 0 \tag{25}$$

Where the Terms A, B, and C can be defined as Equation 26, Equation 27, and Equation 28.

$$A = \frac{\Gamma}{4\sigma^* \alpha_L^2 P_t^2} \tag{26}$$

$$B = \frac{\bar{k}\Delta H_f}{k_L(1+\bar{k})}\,\xi_L + \frac{m_L C_0(k-1)}{D_0(1-(1-k)Iv(P_C))}\,\xi_C \tag{27}$$

$$B = \frac{\bar{k}\,J_S}{k_L(1+\bar{k})}\,(\xi_L + \xi_S) \tag{28}$$

The velocity produced by the LKT Model in Equation 25 and undercooling from Equation 8 can finally be applied to the CNT to produce a delay time Δt between the primary and secondary phases. In Equation 29, the CNT specifies that the number of nucleations occurring during solidification can be determined from the integral with respect to time of the rate of steady state nucleation J_{SS}. The secondary stable phase is said to have initiated when the number of nucleations $N(\Delta t)$ is equal to 1. If the nucleation rate is known at a given velocity, then the upper bound of the integral Δt required to produce a $N(\Delta t)$ of 1 can be solved for [22].

$$N(\Delta t) = \int_0^{\Delta t} J_{SS} * dt' = 1 \tag{29}$$

The term Δt can be expanded to include velocity and undercooling from LKT in addition to thermophysical properties previously mentioned and the catalytic potency f, Boltzmann constant k_b, dendrite radius R, the Zeldovich factor Γ_z, and the phase activation energy ΔG^* homogeneous. The resulting Equation 30 thereby serves as an estimate of delay time when only growth kinematics and material properties are known [22].

$$\Delta t = \left(\frac{60}{\pi}\,\frac{\Delta H_{f,\delta}}{C_P^L * \Gamma_z * K^+}\right)^{1/4} \left(\frac{R}{\Delta T_\delta\, V^3}\right)^{1/4} \exp\left(\frac{1}{4}\,\frac{\Delta G^*_{hom}*f}{k_b * T_L^\delta}\right) \tag{30}$$

While delays for the compositions listed above were accurately modeled by CNT with the input of LKT based parameters in previous studies, this model fails to account for the effects of convective stirring or any other adjustable parameters. By excluding these parameters, it is inferred that delay time is an uncontrollable process and by extension that microstructure cannot be influenced by external factors.

2.6   Retained Damage Model

An alternative theory into microstructural evolution put forward around the same time as geometric based solidification modeling such as the CNT is known as the Retained Damage Model (RDM). The RDM relies on the thermodynamics of a phase transformation rather than growth geometries or kinematics. It serves as a formalization for how delay time may be calculated from the total available Gibbs free energy ΔG_T required to drive the transformation from metastable to stable phase. Under static fluid flow conditions, this energy quantity can be computed as a sum of Gibbs free energy components each with a known driving force. These can be calculated using parameters and material-specific thermophysical properties. In Equation 31 total free energy ΔG_T can be split into three separate components: ΔG_S, ΔG_M, and ΔG_C. Refer to the nomenclature section for definition of key variables [1].

$$\Delta G_T = \Delta G_S + \Delta G_M + \Delta G_C \tag{31}$$

The first term, ΔG_S in Equation 32, is an expression of the classical thermodynamic driving force due to undercooling between metastable and stable phases [1]. This quantity is defined by stable phase properties from the phase diagram as it relates to stable phase transformation kinetics. The parameters used in this equation will be detailed in Section 2.7.

$$\Delta G_S = \frac{\Delta T_{m/s}\,\Delta H_S}{T_{Ls}\,\Omega_S} \tag{32}$$

The second term, ΔG_M modeled by Equation 33, accounts for energy retained in the metastable protostructure due to primary phase undercooling; the higher the original undercooling, the more damage that is retained and the higher the thermodynamic driving force that promotes subsequent transformation kinetics [1]. This quantity is defined by metastable phase properties (See Section 2.7).

$$\Delta G_M = f_x\,\frac{\Delta T_{m/m}\,\Delta H_m}{T_{Lm}\,\Omega_m} \tag{33}$$

One notable capability of the retained damage model is its adaptability to the inclusion of any additional thermodynamic driving force which might impact phase delay. These additional drivers can include any form of controllable energy sources. The final term of the Gibbs Free Energy equation, ΔG_C shown in Equation 34, accounts for the adjustable quantity of retained damage energy due to convection. The functional relationship is based on the concept that the higher the level of convection in the sample, the higher the driving force that promotes subsequent transformation kinetics. This quantity is defined based on melt shear $\dot{\gamma}$ during primary recalescence [1].

$$\Delta G_C = f(\dot{\gamma}) \tag{34}$$

To obtain an estimate of this convective component, Magnetohydrodynamic (MHD) modeling of internal fluid shear rates based on convection due to internal flow velocities and temperatures is required as will be discussed in further detail in Chapter 4.

A general expression for convective energy as a function of shear rate can be modeled by comparing trends in experimental data obtained from solidification experiments under varying convective conditions. One such tool known as the Read Shockley Model, was inspired by dislocations, a type of defect within a material as it solidifies under induced convection and has been used for interpreting the convective retained damage energy of a material[13]. By following a Read Shockley modeling approach [23], changes in dislocation array tilt angles can be recorded, which reflect the increase in microstructural damage as shear rate is increased with stirring. The form of the Read-Shockley approach shown in Equation 35 dictates that if the free energy added to the system by convection ΔG_C is normalized to material shear rates and plotted against the natural log of those shear rates, a linear trend with a signature slope Δ_m and y-intercept Δ_b unique to a given material composition can be produced. When ΔG_C is plotted as a function of the imposed convective shear rate, the resulting curvature has been shown in literature to closely follow a Read-Shockley fit in stainless steel alloys under low stirring Marangoni-flow conditions common to ground based ESL [13].

$$\Delta G_c = \Delta_m \dot{\gamma} \left[\Delta_b / \Delta_m - ln\dot{\gamma} \right] \tag{35}$$

From this model, convective damage free energy can be shown to increase as shear rate is increased with added stirring. Comparison of high-shear ground-based data to low shear space data suggests that the effect of stirring on convective damage free energy decreases significantly at higher shear values. While the application of a Read-Shockley Model suggests that convective energy reaches its peak during maximum H_{CV} space EML stirring conditions, ground-based EML with greater stirring tends to have lower levels of retained damage free energy than this maximum value. The discrepancy between these data sets is modeled by Read-Shockley as a continuous decrease in free energy as shear rate is increased beyond space-EML conditions. In previous testing of space EML data sets the decrease has been so slight that average convective energy at higher shear rates remains nearly constant.

The Retained Damage Model (RDM) allows for the calculation of the energy quantity in Equation 31 by including the internal convective energy from Equation 35. This total energy quantity approximates the driving force behind a phase transformation when internal stirring is present. Using this quantity allows for a predicted of the delay time between the phases through the implementation of classical nucleation theory [22, 24-36] which states that the delay also known as an incubation time τ between formation of the metastable phase and subsequent transformation to the stable phase, can be evaluated by Equation 36 and Equation 37:

$$\tau = \frac{C_{ext}}{\beta} [\Delta G_T]^{-4} \tag{36}$$

$$C_{ext} = 128 \, \pi \, k_B \, T_{Lm} \, \gamma_{m/s}{}^3 \, f(\theta) \left(\left(\frac{N_A}{\Omega_S} \right)^2 \right) \tag{37}$$

For the ferrous alloys studied, nucleation of the stable phase occurs along the metastable phase subgrain boundaries where the grain boundary energy penalty is minimized. As used in Equation 37, this defines the wetting angle θ (in degrees) from Equation 38 and the geometric function $f(\theta)$ from Equation 39.

$$\theta = \left[\frac{180}{\pi} \, \cos^{-1} \left(\frac{\gamma_{m/m}}{2 * \gamma_{m/s}} \right) \right] \tag{38}$$

$$f(\theta) = \frac{1}{2} \, (1 - \cos(\theta))^2 \, (2 + \cos(\theta)) \tag{39}$$

Using these relationships, previous work defined the wetting angle θ as 20.4° and 33.9° [37], and C_{ext} as 2.20×10^{37} and 3.94×10^{37} atom·J^4/m^{12}, based on a β of 4×10^7 and 2×10^8 atoms/sec, for FeCrNi and FeCo [1], respectively.

The utility of the Retained Damage Model in the calculation of delay time for solidification over classical nucleation theory [22] lies in its ability to not only produce estimates of delay behavior for a given set of experimental conditions, but to then provide a means of estimating how the addition or removal of convective energy can influence those predicted delays. Predictions can be attained for a variety of convective conditions, allowing for a selection of the convective condition that provides an ideal delay time for given experimental conditions. In industry This translates to the ability to directly alter microstructure by introducing a magnetic field of quantifiable strength if the predicted delays and known microstructure associated with those delays are undesirable.

2.7   Thermophysical Properties

Thermophysical properties required for the temperature dependent energy properties ΔG_S and ΔG_M can be found in the literature. In later chapters, an investigation into the convective effects on solidification for two alloys, FeCrNi and FeCo will be performed. Their respective properties have been identified and are displayed in TABLE I [37], and TABLE II [1, 38]. Applied heater pulses of varying duration and magnitude taken during ISS-EML testing can be used to generate thermophysical property data where it is absent in the literature.

TABLE I: Thermophysical Properties of FeCrNi

Variable	Property	Units	Fe60-Cr21-Ni19 at%
Ω_s	γ Molar volume	[m^3/mol]	$7.48 \cdot 10^{-6}$ [37]
ΔH_s	γ Heat of fusion	[J/mol]	11235 [37]
T_{Ls}	γ Liquidus	[K]	1713 [37]
$\gamma_{m/s}$	Surface Energy ($\delta \rightarrow \gamma$)	[J/m^2]	0.40 [37]
Ω_m	δ Molar volume	[m^3/mol]	$7.61 \cdot 10^{-6}$ [37]
ΔH_m	δ Heat of fusion	[J/mol]	10629 [37]
T_{Lm}	δ Liquidus	[K]	1668 [37]
$\gamma_{m/m}$	Surface Energy ($\delta \rightarrow \delta$)	[J/m^2]	0.75 [37]
$\Delta T_{m/s}$	Undercooling ($\delta \rightarrow \gamma$)	[K]	44.2 [37]

TABLE II: Thermophysical Properties of FeCo

Variable	Property	Units	Fe60-Co40 at%
Ω_s	γ Molar volume	[m³/mol]	7.40·10⁻⁶ [1]
ΔH_s	γ Heat of fusion	[J/mol]	14083 [1]
T_{Ls}	γ Liquidus	[K]	1757 [1,38]
$\gamma_{m/s}$	Surface Energy (δ → γ)	[J/m²]	0.206 [38]
Ω_m	δ Molar volume	[m³/mol]	7.40·10⁻⁶ [1]
ΔH_m	δ Heat of fusion	[J/mol]	10767 [1]
T_{Lm}	δ Liquidus	[K]	1733 [1,38]
$\gamma_{m/m}$	Surface Energy (δ → δ)	[J/m²]	0.319 [38]
$\Delta T_{m/s}$	Undercooling (δ → γ)	[K]	24 [1,38]

In Equation 33, the metastable Gibbs Free Energy was multiplied by a scalar f_x. This represents the fraction of metastable driving force which is retained within the material and used to drive the metastable phase transformation. This parameter is derived in Equation 40, from a dimensionless undercooling parameter relating metastable and stable phase thermophysical properties. Characteristic f_x values for the alloy compositions of FeCrNi and FeCo discussed are 0.897 and 0.40 respectively [1,39].

$$f_x = N_\Xi = \frac{\Delta H_m / T_{Lm}\, \Omega_m}{\Delta H_s / T_{Ls}\, \Omega_s} \tag{40}$$

Properties required for the computation of ΔG_C for these alloys must also be obtained. Convective shear rates can be varied over a wide range during microgravity testing and induced flow and melt shear may be evaluated based on a knowledge of the temperature and applied electromagnetic (EM) power [40,41]. The applied EM power will be represented in shorthand fashion as a function of the applied heater control voltage setting (H_{CV}) as outlined in the references which summarize modeling of the convection conditions within the droplet using MHD modeling. This requires an understanding of the variation of thermophysical properties with temperature and for this work the following properties displayed in Equations 41 Through Equation 45 were used with relationships for density ρ (in kg/m³), viscosity μ (in Pa·s) and conductivity σ_{el} (in S/m) as a function of temperature T (in K) identified.

<u>Density</u>

FeCo [42] $\qquad \rho = (9.44 - 0.00115\, T) \times 10^3$ $\hfill$ (41)

FeCrNi [40] $\quad \rho = (8.209 - 0.00071\, T) \times 10^3$ $\hfill$ (42)

<u>Viscosity</u>

FeCo [43] $\qquad \mu = 10^{(-0.6615 + 251\ T^{-1}) \times 10^{-3}}$ $\hfill$ (43)

FeCrNi [40] $\quad \mu = 10^{(-0.01154 + 11.98\, T^{-1}) \times 10^{-3}}$ $\hfill$ (44)

<u>Conductivity</u>

FeCo [44] $\qquad \sigma_e = \left(\left((0.0495 * (T - 273.15)) + 56.08 \right) \right)^{-1} \times 10^8$ $\hfill$ (45)

FeCrNi [40] $\quad \sigma_e = 6.63 \times 10^5 + 380(T - 1713)$ $\hfill$ (46)

Using the properties listed in this section allows for modeling of internal convection through MHD modeling which can then be applied to convection based RDM grounded theoretical applications.

Chapter 3. Methodology

3.1 Overview

In addition to a verified set of thermodynamic parameters and the well-supported grounding theory of the RDM, an experimental procedure is required to obtain conclusive results on the nature of solidifying alloys and the influence that convection may have over their behavior. It is critical that any experimentation conducted to this end occurs in a facility which is isolated from external forces or factors that may impact results. Such a facility must also be equipped to allow for the manipulation of controllable experimental parameters, and the capture of data streams such as pyrometry and high-speed video discussed in Chapter 2. Real-time decision making over direct experimental control should be performed by an experienced staff at a remote location that is equipped with proper communication tools between the control room and experimentation facility. Safety of personnel during the experiment must be held as an utmost priority with preventative measures in place. For solidification studies, there are two known experimental setups which meet these requirements. They are the Electrostatic Levitator (ESL) and the Electromagnetic Levitator (EML) and they each allow for the direct isolation of a small spherical metal sample from its external environment by levitating it in place inside of a small enclosure. Both EML and ESL equipment can be used either in space or on the ground with varying success. Ground-based methods subject the sample to gravitational forces resisting levitation which must be accounted for through the analysis of internal convective flows as will be discussed in Chapter 4. Conversely space-based methods have a zero-gravity environment where convection is easier to control. This chapter will discuss the experimental setups that were selected for the creation of a data set against which RDM based models could be compared. Sample selection and preparation for two unique ferrous alloy compositions FeCo and FeCrNi will be highlighted, in addition to experimental procedures that have been put in place for operational control of the sample during experimentation. Finally, methods for data analysis will be discussed.

3.2   Experimental Setup

Of the two experimental setups, ESL is commonly used to obtain thermophysical properties such as the ones referenced in Chapter 2. In a typical ESL facility, levitation is achieved by placing a small, light sample between pairs of opposing electrodes. Coulomb forces then act on the sample, pulling it away from one electrode and toward the other [45]. In ground-based setups, only two electrodes are required, with the attracting electrode positioned such that it's pull is against gravity [45]. When all forces are balanced, the sample will levitate in place. Minor lateral instabilities in the sample can disrupt levitation, resulting in an ejection. If this occurs while the sample is molten as is often the case in solidification studies, the sample may fuse to the walls of the experiment chamber. This is to be avoided as the sample will immediately be contaminated by evaporated particles from previous experiments, many of which would come from different alloy compositions being studied. The importance of maintaining compositional integrity of the samples has led to the addition of advanced multi-axis positional monitoring systems which can detect minor instabilities in the sample and alert the operating technician or system that an adjustment in forces is required [45]. In space-based setups such as the Electrostatic Levitation Furnace (ELF) which was installed to the International Space Station in 2016, additional electrodes are required to keep the sample in place. For both ground and space variations of ESL, the sample can be melted with a high-powered laser [45].

EML is preferred for cases where experimental data is collected with the intent of quantifying convection. To study the full range of convective conditions that a material might experience during manufacturing processes, experiments run using both ground and space setups are required. Ground based facilities taking advantage of gravitational stirring are ideal for creating a baseline data set at maximum convection conditions, while space-based facilities operate with only minor convection unless it is manually added. EML systems rely on layered electromagnetic coils channeling electric current of up to 10A at a known control voltage to generate magnetic fields [17, 46]. Samples placed between two EML coils will become fixed in position. A sample pulled by electromagnetic fields with enough force will eventually melt, allowing groups of coils to selectively heat or levitate samples. Alternatively, lasers can be used in the same manner as ESL. The experimental data set obtained for this book was collected using both the DLR ground-based EML, and the space-based ISS EML which was installed to the Columbus module in 2014 [46].

The ISS EML system operates with a state-of-the-art super positioning (SUPOS) coil which was specifically designed by the German Space Agency (DLR) to handle simultaneous heating and positioning requirements that are independently adjustable [46]. It imposes a weak multi-directional (quadrupole) field at 140 kHz to lock the sample in place between two horizontally aligned coil loops with currents traveling opposite to one another [17]. The fields approach the vertical central axis from all directions in the horizontal plane, then diverge at the exact center between the coils to loop up and over the top coil or down and below the lower coil [46]. A sample caught in the center of the SUPOS coils will be centered by equivalent forces on all sides aligned at 45-degree angles to the horizontal plane. The coils simultaneously produce a much stronger singular direction (dipole) field at 30 kHz along the vertical axis which applies equal force along the equator of the sample perpendicular to the direction of the field capable of heating the sample

up to a rate of 100 K/s [46, 17]. The strength of either field can be augmented by adjusting the

voltage supply to the coils. These voltages are relayed through remote tele-science to the ISS EML

module by operators based on the ground at DLR-Köln in the control room of the Microgravity

User Support Center (MUSC) in the form of a parameter set [46]. A Picture of the MUSC control

room is shown in Figure 2

Figure 2 – MUSC at DLR-Köln

The transmitted file includes specifications for the duration that the voltage will be supplied to

each coil. The utility of this system lies in its ability to rapidly turn heating on or off as desired

without affecting sample position.

Any changes in state during the experiment are monitored in real-time by the axial camera and

pyrometer package (ACP) which is aligned along the vertical axis of the SUPOS coil. The

pyrometer has a sampling frequency of 100 Hz and a wavelength of 1.45 to 1.80 μm and the axial

camera presents a top-down view of the experiment with a maximum resolution of 1280x1024 pixels at 15Hz and can reach a maximum of 200Hz at the cost of resolution which is lowered to 280^2 pixels [17]. For solidification studies, 200Hz is not sufficient to capture recalescence events so a secondary high-speed camera (HSC) is required. This camera is mounted along the horizontal plane and reaches a maximum frame rate of 30 kHz with a resolution of 256^2 pixels [17]. Video data from this camera is temporarily stored on the ISS during the experiment and can be downloaded to the control room servers for review once processing has completed for a given experiment [17].

3.3   Sample Preparation

Two ferrous alloys, FeCo and FeCrNi were selected and processed for delay and undercooling data using the ISS-EML facility [17, 47] onboard the International Space Station. These alloys were selected based on their common use in aerospace applications. Atomic compositions of 60-40 for FeCo and 60-20-19 for FeCrNi were chosen based on availability of ESL collected thermophysical properties in literature. Samples matching the desired atomic compositions for each alloy were ultimately manufactured with tight tolerances alongside several spares. The first sample, Fe60Co40 (at%), was prepared at the German Space Agency (DLR) in collaboration with Thomas Volkmann and Olga Shuleshova with tolerance requirements of ± 1 at%. Composition of the stock material (99.995% Fe rods, and 99.99% Co rods) was validated by energy dispersive X-ray spectroscopy (EDX) prior to manufacture. An example of an EDX view of a sample is shown below in Figure 3

Figure 3 – View of a Post EML Run Sample Surface Through EDX

The desired sample diameter required for ISS EML is 6.0 ± 0.3 mm. Samples were produced in bulk, and any that did not meet this requirement were discarded. To manufacture an EML sample, stock materials are first cut into chunks of relative size then cleaned of debris with ethyl alcohol and dried before they are measured on a microbalance for correct composition. Individual components must be joined into a single sample through the process of arc melting, whereby two electrodes are positioned over measured components under vacuum conditions of under 10^{-5} mbar with 5N Argon gas backfill. A getter material is placed alongside the components to attract any oxygen which might remain in the chamber after it has been vacuum pumped, making sure that excess oxygen does not enter the sample during forming. The electrodes were activated three times on each sample around the seam between the components to ensure proper melting. This process is followed by sample massing to confirm that evaporation loss during arc melting was insignificant enough to change the atomic composition. Sample candidates were then cast to a roughly spherical shape inside of a vacuum sealed furnace (also filled with 5N Argon gas) by raising their temp above the liquidus and causing the sample to melt into a mold. To smooth the sample surface, the cast samples were finally placed in a ground based EML chamber (filled with

6N Helium gas) where it the sample is melted several times under the power of the EML field where an even distribution of forces make the sample spherical. The finalized FeCo samples were measured for compositional correctness using inductively coupled plasma with optical emission spectrometry (ICP-OES) and dimensions can be measured with an ASTRIUM gauge to confirm compliance with desired tolerances. Successful samples were stored in Isopropanol until they can be placed in individual sample cages (inside of which they will be transported to the ISS EML and processed). The sample cages are necessary to prevent the sample from floating away from the EML coil when it is deactivated in zero gravity and also allow for safer transport.

The second sample was a stainless-steel Fe60Cr21Ni19 (at%) alloy that was prepared at Ulm University in collaboration with Hans-Jörg Fecht, Rainer Wunderlich, and Markus Mohr [48]. The same process outlined above for the manufacture of FeCo samples was used in the creation of FeCrNi samples with the exception that for this material, a 6.5 ± 0.3mm diameter was selected instead of 6.0mm. The resulting component % weights of the selected EML sample were Fe 60.3%, Cr 19.8%, Ni 19.9% which meets the 1% deviation from composition requirement for flight samples.

3.4   Data Collection Procedure

With the prepared samples successfully loaded into their cages and housed on the sample carousel inside the ISS EML, experimentation can commence. Scheduling is coordinated closely between astronauts and the MUSC ground team to ensure that vibrations from work done on the station will not influence the testing. In preparation for the experiments, the HSC is set to recalescence mode, with a frame rate of 30kHz. The sample chamber where the EML SUPOS coils are contained is then pressurized over an extended period using an inert gas of either Helium or Argon to prevent oxidation of the sample during processing. Delayed video feed from the ISS EML ACP unit is next loaded onto a MUSC control center monitor and the pyrometer is initialized to the current time (GMT). A running plot of temperature at 100Hz is transmitted from the ISS EML ACP to the MUSC experiment display.

Researchers next began by selecting one of the alloy compositions outlined in the previous section. The expected liquidus (melt) temperature for that alloy is then recorded for comparison against the pyrometry feed. This will assist with estimation of the point in time when the metal sample has become fully molten. It is desirable to keep the sample at peak temperatures for only short periods of time because the longer the alloy remains at high temperature, the more likely it will be to evaporate excessively and change compositions. Evaporation is accompanied by the buildup of toxic chemicals on the surface of the sample chamber, the maximum quantity of which is known as the toxicity limit which is strictly limited by NASA to provide safety for the astronauts if the sample chamber is breached and the fumes released. Simulations are performed in advance to estimate both the quantity of evaporation and sample toxicity. Experiment run time is limited to an available dust budget which is split up among the samples within a given batch. The total number of allowable sample tests is determined in advance from the available dust budget and expected toxicity.

Researchers have a choice in the applied heater control voltages and the duration of heating for each run. Care is taken to limit heating duration between from the time when the heater is activated to when it is disabled such that the sample is fully molten at heater shutoff as viewed from the pyrometry feed. The heater can be set to an automatic shutoff at either preset temperatures or preset times. Typically, the desired superheating to some temperature above the expected liquidus is used as the shutoff. However, this temperature must be raised to account for any emissivity shifts, the presence of which will be evident after the first test run if the pyrometry feed at liquidus does not match the expected liquidus. In this case the superheat temperature setting is defined based on observed liquidus.

Researchers must plan whether to pulse the heater with a step function of voltage at some point after heater shutoff while the molten sample is solidifying. Such a pulse must be specified from 0.1V to 5.0V dictating the amount of stirring that will be induced within the sample for that test run. 5.0V is associated with maximum induced stirring while 0.1V is the minimum representing no stirring. Tests are split into groups such that they cover this full range of voltages to ensure adequate coverage of convective shear rates required for the RDM. At least 2 unique voltages must be represented in the final data to produce a trend. The duration of each pulse is independently determined based on the expected time from heater shutoff to recalescence as observed in previous test runs. Leaving a gap between pulse shutoff and recalescence ensures that the temperature decrease will not directly impact the temperature rises produced by recalescence.

With stirring voltages planned for each test run, the sample is loaded into the sample chamber and observed on the ACP video feed. MUSC technicians then remotely upload parameter sets with voltage pulse settings for the first test to the ISS EML. The test is initiated on command. First the positioner coil is activated, pulling the sample off its resting pedestal. Then heating commences until the predefined superheat temperature limit is reached when the sample is fully molten. Upon heater shutoff the sample will begin to free cool, at which point the heater will pulse on to the predefined voltage and stay on throughout the cooling process to induce stirring. The pulse will be disabled automatically prior to recalescence. The sample will finally be allowed to free cool having attained the crystal structure of the final phase transformation that it underwent during recalescence. The positioner remains on while Researchers wait for the facility to download the HSC video transmitted from the ISS EML. Upon video review, it is observed whether the phase transformation initiated on the side of the sample facing the HSC. Data can only be collected if the transformation is visible. If the transformation is not apparent on the HSC, the test will be repeated until successful. Each successful run will yield a single data point for delay time and undercooling.

3.5   Computing Delay Time

Investigating video footage downloaded from the ISS EML HSC allows for frame-by-frame observations of phase transformation events occurring on the surface of the FeCo sample. As the primary or secondary phase grows across the sample surface, selections of key edge points were made using a MATLAB program (See Appendix A). In the mosaic of Figure 4, sequential high speed video frames of a FeCo sample processed at low undercooling and without stirring depict primary and secondary phase progression. In each frame, colored point selections along the phase edge have been superimposed onto the sample image. They mark a clear color change between the lighter phase growth and the darker surrounding fluid. The first four frames show the primary metastable phase growing into undercooled liquid, while the second set of four frames represent the formation of a secondary stable phase growing into the metastable mushy zone. In the second set, visible white dots represent oxide tracer particles (higher emissivity than the liquid) floating on the surface of the droplet.

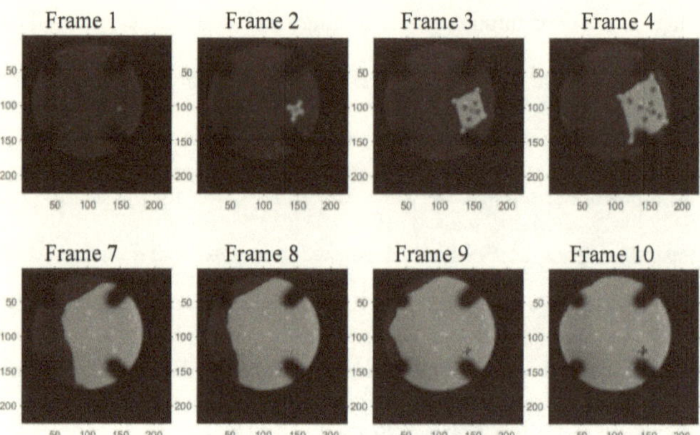

*Figure 4 - **Mosaic of Radial Camera Generated Video Frame Images***

X-Y coordinates for each point selection relative to the upper left-hand corner of the frame are recorded. An approximation of the initiation point for both the primary and secondary phases can be triangulated from the selected points in a given frame close to the time of nucleation. Then, by assuming a stationary ellipsoidal shape with a known radius and center coordinate from the observed sample profile, the Euclidean distances between each edge point in the first frame of a phase and phase center can be calculated. The distances between all points in a given frame are averaged to obtain an average growth distance between nucleation and that frame [39, 49]. Average growth distances between any two subsequent frames can similarly be calculated from the average Euclidean distances between each point's original and final positions.

For each phase edge, the cumulative distance from the phase center at any given frame in the video can be taken as the sum of that points Euclidean distances in all previous frame steps. For each frame in the video, the average cumulative distance from phase center of each edge point is recorded. Growth rates are attained by dividing the average cumulative distance by the timestep between video frames. Performing a linear regression on the average phase growth rate yields a prediction of the intraframe phase initiation time. Using this technique, nucleation times occurring between successive frames may be evaluated to obtain continuous resolution during acquisition instead of relying on a discrete frame-by-frame rough approximation. If two phases form during the same cooling cycle, the duration between their phase initiation points produces a delay time in units of a fractional portion of a frame. This is the same delay time which can be predicted by CNT or RDM methods. An example is shown in Figure 5, where two intersection points with the time-axis have been identified from linear regression of the cumulative distances at each video frame. The growth rates of the primary metastable phase (Blue line) and secondary stable phase (Red line) are measured in pixels per second. The points in this plot represent the cumulative Euclidean distance of each tracked edge point on a given video frame from the identifiable phase center point on an assumed spherical sample. The delay time for this cycle, recorded as the distance between these intra-frame intercepts, was determined to be 1.096×10^{-4} seconds. When repeated for multiple cycles, a generalized trend can be produced. Tracking delay time as a function of undercooling collected from pyrometry can provide additional insights into material behavior.

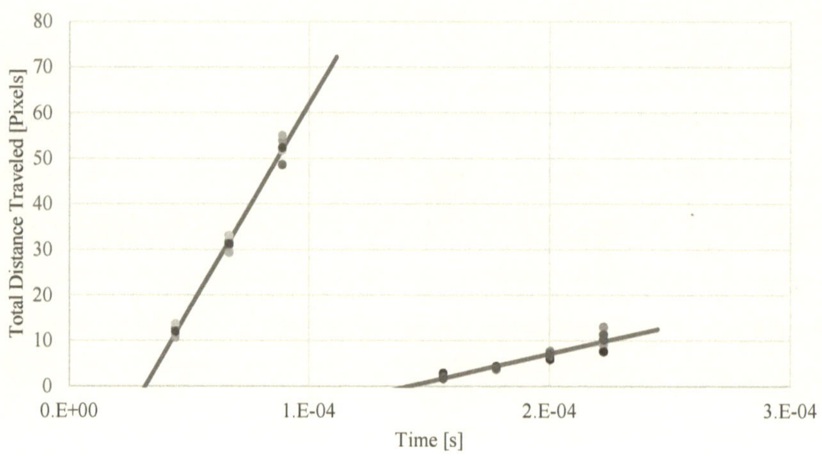

*Figure 5 - **Growth Velocity Plot Showing Two Linear Regressions***

Using the methods outlined above, critical data for sample transformation kinetics was obtained from the high-speed video for both FeCo and FeCrNi, thus defining their experimental delay times. The produced delays are notably independent of undercoolings determined by performing the process outlined in Chapter 2 on temperature data collected from the ACP. Delay times and undercoolings could now be compared to previously collected ground-based data from the DLR EML allowing for functional relationships to be modeled using RDM based methods.

Chapter 4. Modeling

4.1 Overview

Modeling of convective influences on solidification behavior as mentioned in Chapter 2 is dependent on an accurate understanding of internal convective flows occurring at the instant when recalescence is captured on high-speed video and pyrometry. Combining MHD surrogate modeling with RDM theory allows for a better understanding of experimentally obtained observations which is why this is a critical step in any solidification analysis. The functional relationship between flow velocity, shear rate and fluid temperature can be better understood as well. This chapter will cover the MHD surrogate model created for the unique FeCo and FeCrNi compositions discussed in Chapter 3. The resulting estimates for internal convection will then be applied to Read Shockley and Saturation Model approaches to the RDM to compare the effects of temperature on convection. Then in Section 4.3, an optimized saturation approach will be applied to predict delay times for a wide range of undercooling and convective conditions.

4.2 Surrogate Modeling (MHD)

Shear rates within the material must be estimated to provide a model for convective behavior that can complement the delay times collected in a manner outlined in Chapter 3. For these rates to be accurately generated, flow modeling is required over the range of conditions accessible during space testing. MHD simulations for applied heater control voltage H_{CV} driven flows were run for a series of thermal hold conditions both above and below the stable phase liquidus temperature with variable applied H_{CV} from 0.1 V to 5.5 V [40]. Maximum flow velocity and maximum shear rates as a function of temperature and control setting were obtained for a representative nominal space sample size of 6.5 millimeters diameter. A double linear regression was performed on the resulting plot of shear rate as a function of flow velocity and correlated to temperature ΔT above (positive and superheated) or below (negative and undercooled) the stable phase liquidus temperature and simultaneously correlated to the applied H_{CV}. This produces a three-term master equation linking behavior for both FeCo and FeCrNi models. The three master coefficients A, B, and C in a third-order secondary polynomial fit are shown in Equation 47 Through Equation 49 with secondary temperature coefficients d, e, f, and g used to quantify the influence of temperature on the shear coefficient. Equation 50 presents the format for display of the results as a function of velocity (V) and the master coefficients to describe the influence of the heater setting on melt shear coefficients through velocity.

$$A = d_A \Delta T^3 + e_A \Delta T^2 + f_A \Delta T + g_A \tag{47}$$

$$B = d_B \Delta T^3 + e_B \Delta T^2 + f_B \Delta T + g_B \tag{48}$$

$$C = d_C \Delta T^3 + e_C \Delta T^2 + f_C \Delta T + g_C \tag{49}$$

$$\dot{\gamma}_{max} = A\, V^2 + B\,V + C \tag{50}$$

The secondary coefficients are displayed in TABLE III for this equation for both FeCo and FeCrNi. Using Equation 50 with appropriate coefficients for the selected material and hold temperature allows for the calculation of a maximum shear rate to match against delay times when only applied H_{CV} (in volts) and superheat/undercooled melt temperature ΔT (in K) are known.

TABLE III: Turbulent Shear Rate Secondary Coefficients

Coefficient	$d\ (\Delta T^3)$	$e\ (\Delta T^2)$	$f\ (\Delta T^1)$	$g\ (\Delta T^0)$
A_{FeCrNi}	3E-5	0.013	1.911	150.36
B_{FeCrNi}	-7E-6	-0.0033	-0.7371	1849.8
C_{FeCrNi}	-3E-8	2E-5	0.0151	-6.2478
A_{FeCo}	0	0	0	0
B_{FeCo}	-3E-7	-0.0002	0.0243	1922.6
C_{FeCo}	0	5E-6	-0.0046	-7.8367

4.2.1 Velocity Change with HCV - Observed stirring Effect on Internal Flow.

In Figure 6 MHD modeling using the thermophysical properties of FeCo was conducted for a series of incremental thermal holds spanning the range of experimentally observed sample undercoolings reaching down to T_m-300 (and sample superheat conditions up to T_m+300) [40]. For each thermal hold represented by a colored line, the H_{CV} responsible for stirring was adjusted from a minimum 0.10 V, representing no induced stirring, through to a maximum 5.50 V representing maximum EML space stirring condition. Maximum turbulent velocity for each simulated H_{CV} stirring setting at a given thermal hold has been recorded. The resulting plot may be readily compared to similar representations in the literature as performed on FeCrNi [40]; the older results and the newly presented results are virtually indistinguishable leading to confidence in the master equation approach.

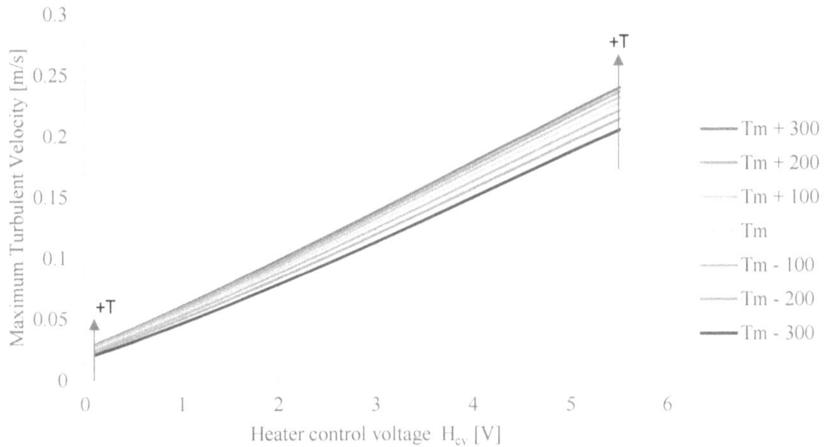

Figure 6 - FeCo Turbulent Velocity as a Function of H_{CV}

4.2.2 Shear Rate Change with Velocity

Next, to identify variation between FeCo and FeCrNi, expected internal fluid shear rates of the sample were obtained from the same MHD simulation used to sample velocities. Correlations were then made to three key master-equation dependencies: temperature, H_{CV}, and maximum flow velocity. This was done for both FeCo in Figure 7, and FeCrNi in Figure 8 assuming turbulent conditions. Experimentally, lower H_{CV} settings yielded laminar flows, which were also modeled. Laminar correlations are of similar composition and as such are not represented in the figures below. Each thermal hold is shown to span a variable range of maximum velocities. Colored lines for each thermal hold initiate at a low H_{CV} (0.10 V) marked by a blue dot and terminate at high H_{CV} (5.50 V) marked by a red dot. As hold temperature is progressively increased, conditions at maximum and minimum H_{CV} migrate toward higher shear rates as represented by blue and red arrows.

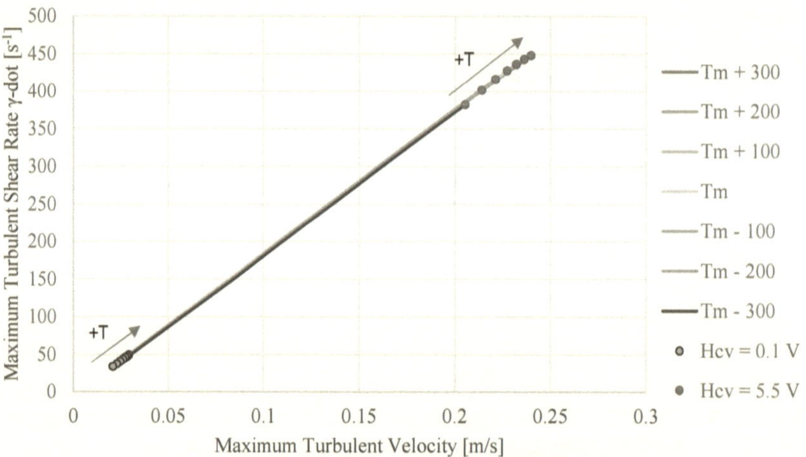

*Figure 7 - **FeCo Shear Rate vs. Velocity (Turbulent)***

42

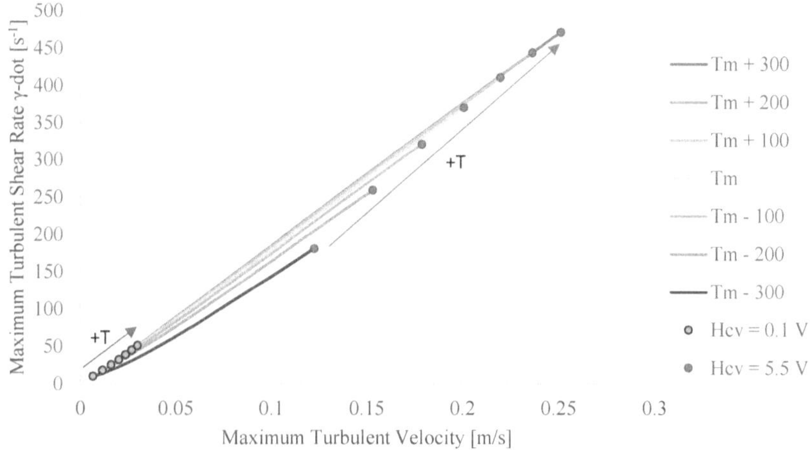

*Figure 8 - **FeCrNi Shear Rate vs. Velocity (Turbulent)***

4.2.3 Convective Free Energy Change with Shear Rate

Using MHD modeling, shear rates were produced for the known undercooling and flow velocity conditions of the experimental data set. Normalizing the experimental ΔG_C data to shear rate and plotting as a function of the natural log of shear rate produced an approximately linear trend which was defined by parameters for slope Δ_m and y-intercept Δ_b [2]. A Read-Shockley Model using these input parameters could be implemented and has been recorded as a green curve on all subsequent plots in this chapter. It will serve as a reference for the new model which will be developed in this section.

An alternative approach to the Read Shockley Model based on a saturation approach [50], was selected for comparison to approximate the average convective energy as a limit such that the effect of shear rate has a positive effect on convective energy across the entire range of possible shear rates. The resulting Saturation Model is shown in Equation 51 to have a slightly different structure than the Read Shockley Model presented in Chapter 2.

$$\Delta G_c = a\,\{1 - \exp[-b\,\dot{\gamma}]\} \tag{51}$$

At low shear rates typical of low-H_{CV} space-EML testing, the Saturation Model begins by increasing convective energy at a constant rate. Increasing parameter 'b' will produce a steeper slope. Slopes created through least-square fitting are similar to those of the Read-Shockley Model under similar conditions. Increases in convective energy then gradually taper off toward a clear limit at higher shears, characterized by parameter 'a'. The parameter 'a' may be defined such that it asymptotically approaches the average convective energy at moderately high shear modeled by Read-Shockley. However, unlike the Read-Shockley Model, there is no subsequent decrease in convective energy at high shear using this model.

Accuracy of the two models can be evaluated based on the deviation of experimental results from model predictions. The magnitude of this deviation was significant between the two ferrous alloys studied. For this model, fit parameters *'a'* and *'b'* were obtained through least-square fitting to experimental data yielding constants for initial incline and plateau behavior for the saturation curve model for both FeCo and FeCrNi alloys as shown in TABLE IV. The slope and y-intercept of Read-Shockley curves were also obtained and are likewise included.

TABLE IV: Model Curve Fit Parameters

	Read Shockley Model		Saturation Model	
Composition	Slope (Δ_m)	Y-intercept (Δ_b)	Initial Incline (b)	Plateau (a)
FeCrNi	-250321	1681683	0.0149096	$70125700 \pm 1.44e7$
FeCo	-571290	3677220	0.0565790	$90651200 \pm 2.28e7$

With shear rates from MHD modeling recorded for each observed instance of space and ground EML data, ΔG_C can be computed using either Read-Shockley or Saturation Models using the parameters in TABLE IV along with Equation 35 and Equation 51 then plotted as a function of calculated shear rate. Several plots have been constructed to highlight the dependency of shear rate on flow conditions during phase transformation events. This is a necessary addition to the study as convective energy within the sample, and thus solidification behavior as a whole, is largely dependent on the properties and material response of shear flows to the addition of stirring.

Leveraging shear rates obtained from MHD surrogate modeling for known temperatures and H_{CV} induced velocities of the experimental FeCrNi and FeCo data sets, RDM-enabled convective free energy models provide access to trends in convective energy relative to induced sample stirring. In both Figure 9 and Figure 10, Saturation and Read-Shockley Model-generated predictions for the available quantity of ΔG_C at a given $\dot{\gamma}$ are shown by Blue and Green lines respectively. Each models' predictions have been superimposed onto a scatter plot of experimental space data collected over the full range of space-based stirring conditions, with H_{CV} levels ranging from 0.10 V through to 5.5 V. Data collected from ISS-EML testing with applied stirring is represented by color-filled symbols specific to the applied H_{CV} setting, while ground based EML data without applied stirring is shown by open symbols. As the heater control voltage is increased to induce more stirring, convective free energy and shear rate both steadily increase. EML space data of this form is characterized by shear rates less than $\dot{\gamma} = 350\ s^{-1}$. Each model is grounded by tight clusters of experimental ground based EML datapoints (in gray) centered- in bulk on $\dot{\gamma} = 450\ s^{-1}$. In both FeCo and FeCrNi, the magnitude of the Saturation Model's (Blue line) limiting parameter 'a' is evident as it reaches a plateau in convective free energy in the region populated by data above an applied $H_{CV} = 2.00$ V. In Figure 9 the Read-Shockley curve (Green Line), also fit to the FeCo data set overshoots the limit set by the Saturation Model for that alloy, then curves back down indicating an unnatural reduction in convective free energy at higher shear rates. Yet in Figure 10, the bulk of ground-based data (open symbols) have a similar convective free energy to space EML data with the highest amount of stirring at $H_{CV} = 4.40$ V. Because there is no notable decrease in convective free energy with increases in shear rate, both the saturation curve and Read-Shockley curve more closely match one another in FeCrNi than the same plot for FeCo.

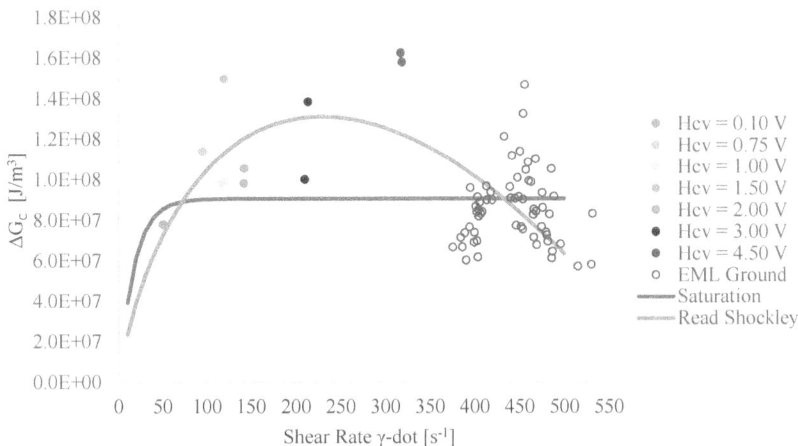

*Figure 9 - **FeCo Plot of Convective Free Energy vs. Shear Rate***

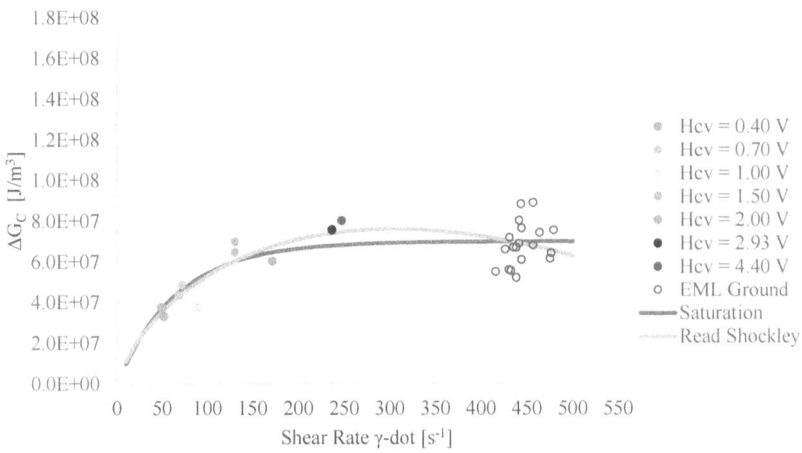

*Figure 10 - **FeCrNi Plot of Convective Free Energy vs. Shear Rate***

Normalization of ΔG_C by $\dot{\gamma}$ and plotting against the natural log of $\dot{\gamma}$ for the same set of experimental space and ground EML data, as done in Figure 11 and Figure 12 for FeCo and FeCrNi respectively allows for better visualization of the Read-Shockley (Green line) linear slope and y-intercept parameters. For FeCo the Read-Shockley curve is shown to be linear in nature, with an anchoring point in the center of the gray ground-based data point cluster. The space EML data (colored points) have a wide spread to either side of the linear trend. In general, data at higher stirring H_{CV} (1.00 V to 4.50 V) was below Read-Shockley curve, while data collected at less than (1.00 V) is above the Read Shockley curve. The saturation curve (Blue Line) more closely follows this observed curvature while still anchoring to the ground-based data. For FeCrNi all data closely follows the linear path of the Read-Shockley trend line, over the range H_{CV} = 0.40 V to 4.40 V, with each stirring level differentiated by symbol color, and high stirring gravity driven EML ground data (open symbol). The closeness of fit forces the normalized saturation curve to follow a near linear path.

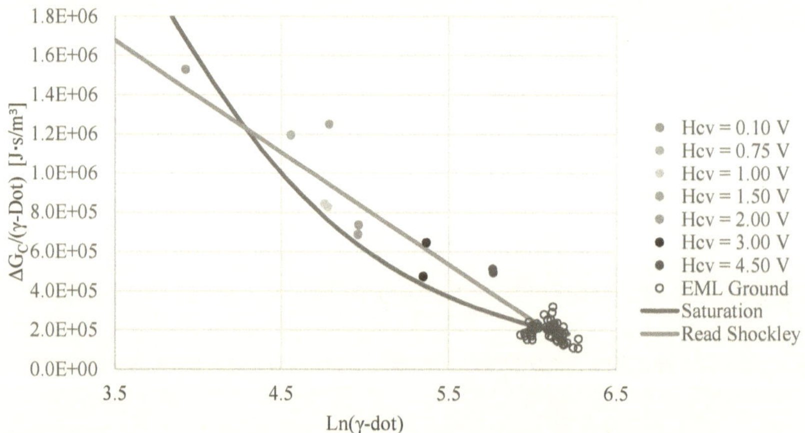

Figure 11 - FeCo Plot of Convective Free Energy Normalized by Shear Rate

48

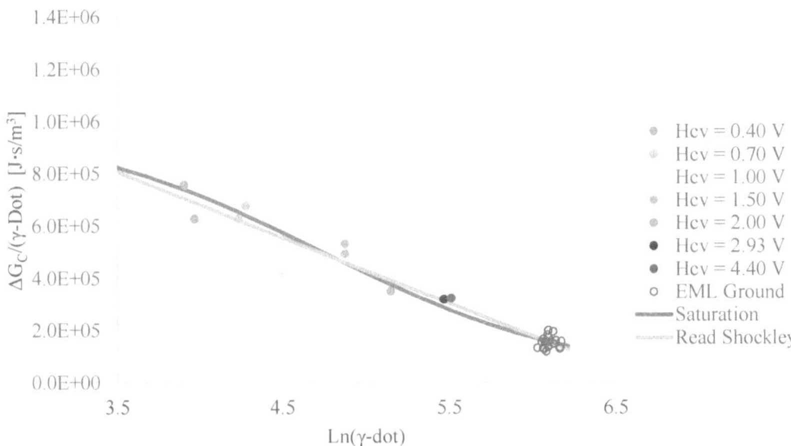

Figure 12 - FeCrNi Plot of Convective Free Energy Normalized by Shear Rate

4.3 Modeling the Influence of Convection by a Saturation Approach

Following the theory outlined by the RDM, a saturation approach to convective energy approximation was ultimately selected for the prediction of delay times across a range of undercooling and convective conditions using Equation 36. This is contrary to previous work where a Read Shockley approach was favored [1, 13, 37]. Delay predictions computed from a Saturation Model could be plotted as a function of undercooling against the delay times for newly attained FeCo ISS-EML experimental data and ground-based data taken for the same alloy composition at peak shear conditions.

4.3.1 Saturation Delay Prediction

Solidification delay time predictions were computed for both FeCo and FeCrNi. The results of these predictions can be found in Figures 13 and 14. In each plot, the same experimental EML data set used in Figure 9, Figure 10, Figure 11, and Figure 12 has been represented by delay time τ and observed undercooling ΔT (instead of the previous ΔG_C and $\dot{\gamma}$). Once again, data covers a range from H_{CV} = 0.10 V to 4.50 V with each stirring level differentiated by symbol color. In this representation, ground based EML data points (still Gray) were not manually stirred, yet the gravitational forces present induced a level of stirring greater than the H_{CV} = 4.50 V space test. They are no longer clustered together and span the entire range of possible delay times. Delay time is shown to decrease with increasing undercooling and stirring. For two samples with equivalent undercooling, yet different stirring conditions, the sample with higher stirring will have a lower delay time. Data therefore approximates a range of delays for any undercooling. To estimate this range, values of shear rates at both high and low stirring conditions were substituted into the saturation curve predicting delay time at a given undercooling.

For FeCo, experimental shear rates are characterized by $50\ s^{-1} < \dot{\gamma} < 400\ s^{-1}$ These predictions diverge from the data set at low undercooling. FeCrNi has a range $50\ s^{-1} < \dot{\gamma} < 500\ s^{-1}$. Saturation Model prediction lines are created by solving for the reference delay time in Equation 36 while varying $\Delta T_{m/m}$ in the ΔG_M term and using a substitution of ΔG_C at a constant $\dot{\gamma}$ from Equation 51. A saturation line was constructed for both minimum (Blue) and maximum (Red) shear rate conditions for the specified material. Each line predicts τ for a given ΔT at the specified shear rate. For FeCo, data is clustered around the saturation predictions but is not bounded by either upper or lower predictions. FeCrNi data has a better fit to the saturation curve during modeling and mostly bounded within the prediction ranges. In that alloy, Low H_{CV} data (orange, yellow points) accurately follow the Blue prediction, while higher H_{CV} (Light Blue, Blue, Brown) data follows the Red prediction as expected.

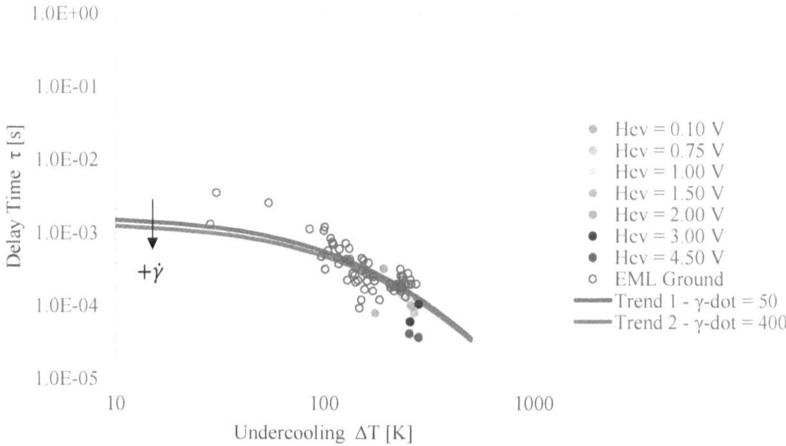

*Figure 13 - **FeCo Saturation Predictions for Delay Time vs. Undercooling***

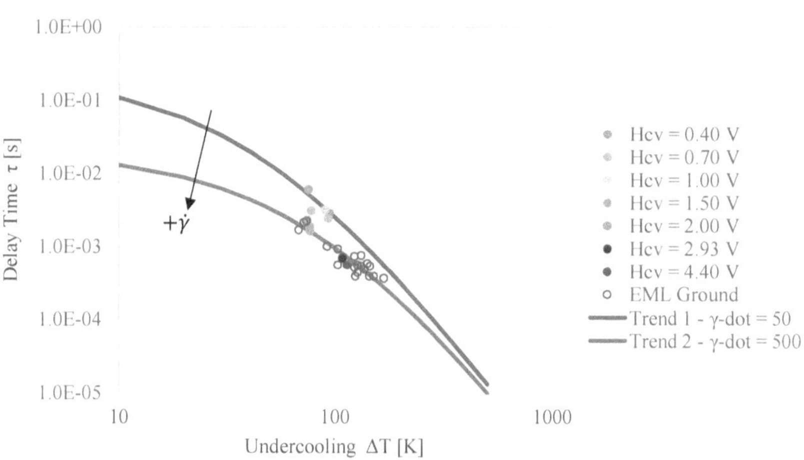

*Figure 14 - **FeCrNi Saturation Predictions for Delay Time vs. Undercooling***

4.4   Model Validation by Dimensionless Predictions

The performance of model predictions for delay time can be evaluated by checking if the relationship between independent dimensionless parameters adhere to the expected behavior of the RDM [1, 14]. For Saturation and Read Shockley Models, two such parameters exist. The first parameter is that of the nondimensional thermal driving force N_M which can be evaluated by comparing calculated ΔG_T (Obtained using experimental undercooling data for ΔG_M in conjunction with Saturation or Read Shockley approach for ΔG_C) against the smallest possible available free energy in the system required to drive the transformation during solidification. In the case of solidification, he minimum energy quantity occurs when there is no primary phase undercooling $\Delta G_M = 0$, and the sample is quiescent with no internal flow velocity $\Delta G_C = 0$. Under these conditions only ΔG_S is present as computed in Equation 32 using thermophysical properties of the stable phase. It can be referred to as the reference state ΔG_R, where $\Delta G_R = \Delta G_S$. The relation N_M between total energy under experimental conditions ΔG_T and the energy at the reference state ΔG_R is dependent only on the experimental undercooling as collected from radiation pyrometry and on the convective properties of melt shear at that undercooling. As such, ΔG_R is independent of delay data collected from the high-speed camera [1, 14]. The Equation for N_M has been presented below in Equation (52).

$$N_M = \frac{\Delta G_T}{\Delta G_R} = \frac{\Delta G_T}{\Delta G_S} \tag{52}$$

The second dimensionless parameter is conversely independent of undercooling and dependent only on delay times under experimental conditions as captured on high-speed video. The selected reference state required for this parameter matches the conditions of ΔG_R, where the shortest possible driving energy quantity produces the longest possible delay [1, 14]. This reference delay

τ_R can be computed by Equation 9 where $\Delta G_T = \Delta G_R = \Delta G_S$. The ratio between experimental delay and reference delay produces the quantity N_D which is defined by Equation (53).

$$N_D = \frac{\tau_{exp.}}{\tau_R} = \frac{\tau_{exp.}}{\frac{c_{ext}}{\beta}[\Delta G_R]^{-4}} = \frac{\tau_{exp.}}{\frac{c_{ext}}{\beta}[\Delta G_S]^{-4}} \tag{53}$$

Because dimensionless delay is a function of the reference state energy, ΔG_R, it can be said to vary with dimensionless driving force. Evaluating N_M vs. N_D on a Log-Log plot will produce a linear trend where delay decreases as driving force is increased as shown in Figure 15 for the FeCrNi alloy using a Read Shockley approach to computing ΔG_T [14]. Points are separated into three groups including 0-volt ISS EML space data in Blue, stirred ISS EML space data in Orange, and gravity stirred EML Ground data as Hollow data points. If the slope obtained by linear regression is sufficiently close to m = -4, mirroring the relationship between N_M and ΔG_R shown in Equation 53, then the independent measures of delay and undercooling agree with one another. Any agreement also will indicate that the Saturation or Read Shockley Model for delay time created using an estimate of thermodynamic driving force ΔG_T from this data set will accurately predict the delays based on RDM theory. In Figure 15 the identified slope for FeCrNi using the Read Shockley method is -4.00508 which is in close agreement with previous results of -3.938 obtained from Read Shockley modeling of ESL ground data in the literature [1]. Figure 16 represents the same experimental data in a Saturation Model. It has a similarly good fit that is close to m = -4. Read Shockley and Saturation modeling was also performed on the FeCrNi alloy in Figure 17 and Figure 18 respectively. Each adheres closely to the expected slope shown as a black line.

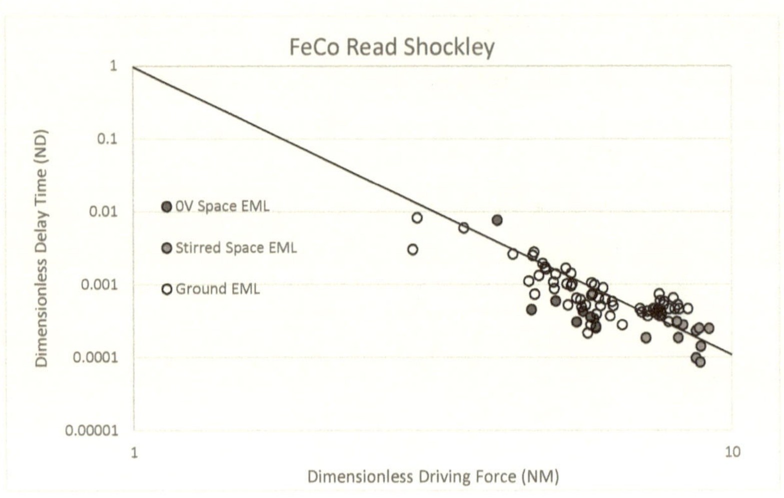

Figure 15 – FeCo Read Schockley Dimensionless Driving Force Plot

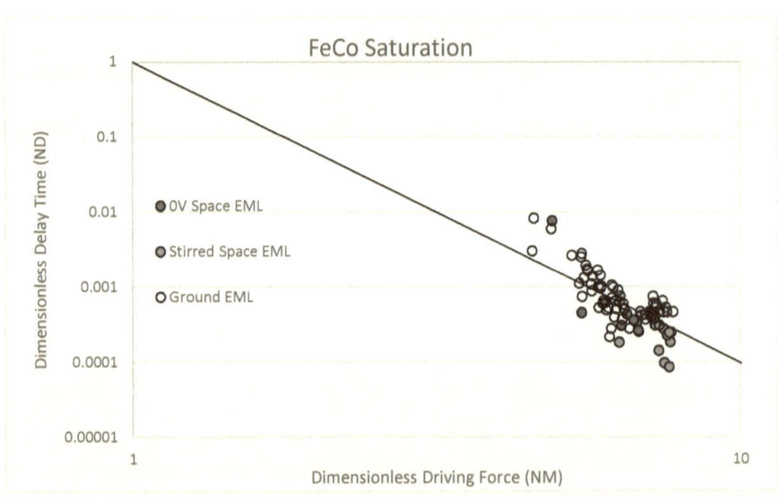

Figure 16 – FeCo Saturation Dimensionless Driving Force Plot

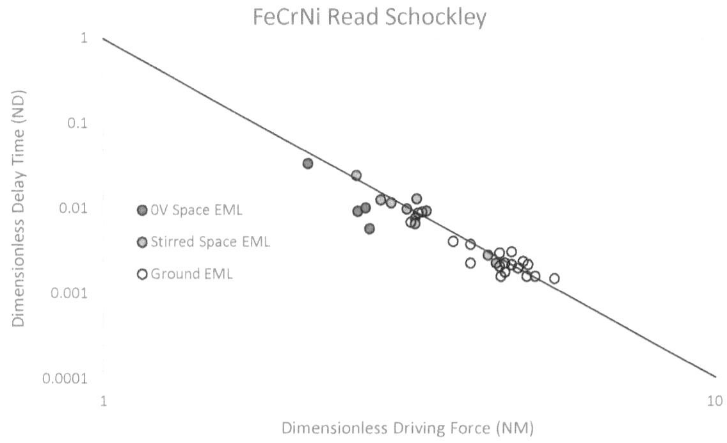

Figure 17 - FeCrNi Read Schockley Dimensionless Driving Force Plot

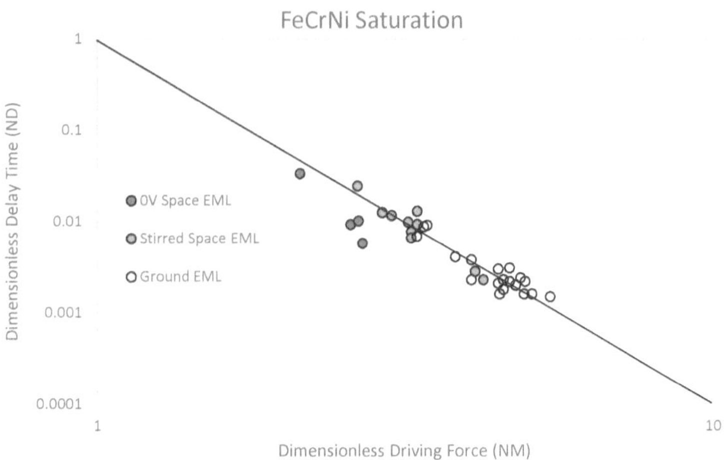

Figure 18 - FeCrNi Saturation Dimensionless Driving Force Plot

This process was repeated for FeCo and the resulting data for N_M and N_D was added to the plot in Figure 19. A reference line in black was added to represent a slope of -4 going through the origin representing 0 delay when no driving force is present. Stirred ISS EML space and stirred EML Ground data sets have been combined for both alloys to provide clearer distinction between stirred (dark colored) and unstirred (bright colored) data. FeCo is in Orange while FeCrNi is in Blue.

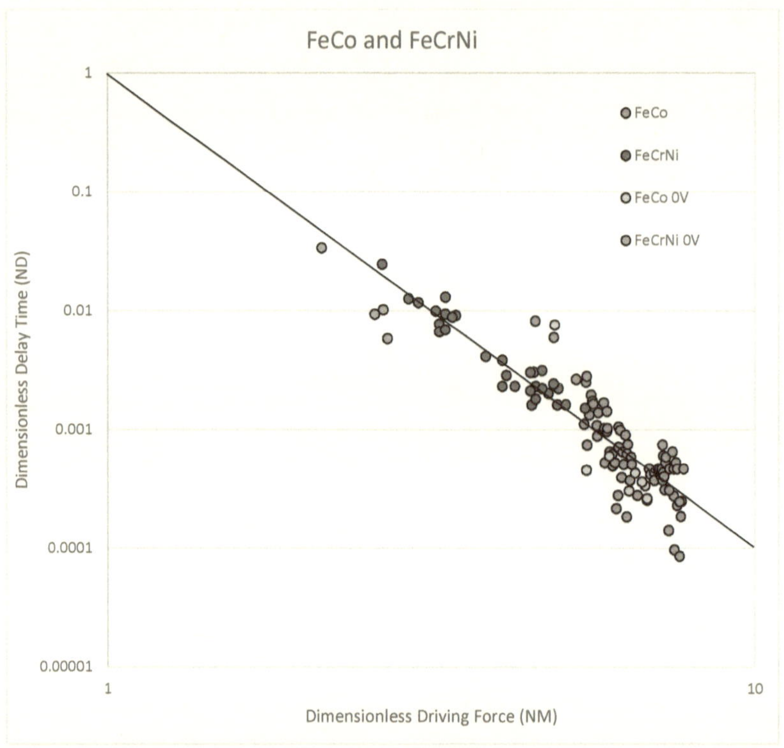

Figure 19 - FeCo and FeCrNi Dimensionless Driving Force Plot

For the stirred data sets obtained using the Saturation Model (dark colored points) a linear regression grounded in point (1,1) was performed. The resulting values for slope are shown in TABLE V. The slope using Read Shockley Methods was similarly obtained and recorded in the table. Uncertainty was calculated for all estimates. A comparison between previously published

FeCo results using Read Shockley methods on an ESL dataset collected from Marshall Space

Flight Center (MSFC) shows good agreement.

TABLE V: Slopes of Nondimensional Delay Time and Driving Force

Composition & Model	Literature Nt/Nm	Calculated Results Nt/Nm
FeCo Read Shockley	-3.938 ± 0.123 [1]	-4.00508 ± 0.033667
FeCo Saturation	-	-3.97953 ± 0.032794
FeCrNi Read Shockley	-	-3.99769 ± 0.031139
FeCrNi Saturation	-	-3.98688 ± 0.031081

4.4.1   Phase Growth Velocity Comparison Against Prior Data

Another form of validation can be achieved through a comparison of growth rate data, which was

collected during the experimental calculation of delay times, against similar data recorded and

validated in past experiments. Looking back to Chapter 2 where the calculation of delay time was

first discussed, the nucleation time of each phase was determined by tracking the visible phase

front position and extrapolating their growth back in time to when nucleation must have been

initiated. The slope of the regression lines that were generated represent the growth velocity for a

specific phase. Knowledge of the delay time and the growth rates allows predictions on phase

selection under conditions of growth competition as described in the literature. [15, 38, 51, 52]

In the case of the current space data for FeCo, we observe growth of ferrite into the undercooled

liquid followed by the growth of austenite into the mushy-zone following primary metastable

recalescence. These space electromagnetic levitation results can be compared to values obtained

using ground-based electrostatic levitation as presented in Figure 20 which is adapted from the

literature [38]. The new space EML growth velocities (solid symbols) as a function of

undercooling for two phases are compared to ground-EML results (open symbols) [7]. The growth of metastable ferrite into the liquid (solid squares) agrees quite well with the Rodriguez result (open squares) but the new results for growth into the mushy-zone (solid circles) are seen to indicate a constant velocity of 5.39 ± 0.39 m/s which is significantly higher than that previously measured value of 2.40 ± 0.23 m/s (open circles) [38]. A statistical comparison showing standard deviations is expanded on in TABLE VI.

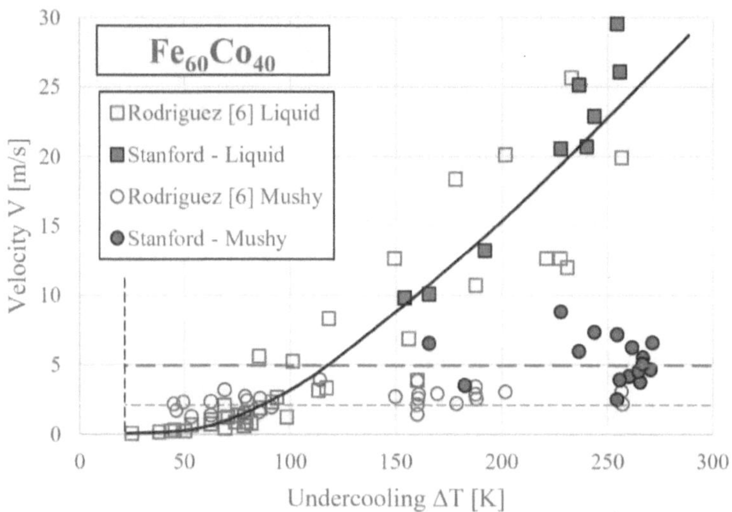

*Figure 20 – **FeCo Metastable and Stable Phase Growth Rates***

TABLE VI: Comparison of Secondary Phase Growth Velocities

Properties	Fe60Co40 Rodriguez	Fe60Co40 Stanford
$V_{\gamma\delta}$ (m/s)	2.40 ± 0.23 [54]	5.39 ± 0.39
Standard Deviation (m/s)	0.64 [54]	1.67
Num $V_{\gamma\delta}$ Data Points	29 [54]	16

Hypothesis testing in the form of an unequal variance T-test was performed on the two data sets provided in this table. A null hypothesis H_0 was first defined such that the means of the two experiments are set equal with H_0: $V_{\gamma\delta \text{ Stanford}} = V_{\gamma\delta \text{ Rodriguez}}$. An alternative hypothesis where the

58

means of the two experiments are taken to be unequal was likewise defined as H_1. The required T-score was computed as $t_0 = 6.898$ with 15 degrees of freedom. At a 5% level of significance, the table returns a value of $t_{\alpha/2} = 2.131$ which is lower than t_0. Because $t_0 > t_{\alpha/2}$ the null hypothesis H_0 can be rejected with 5% uncertainty in favor of the alternative hypothesis H_1. This result suggests that the two populations of collected secondary phase data are significantly different.

4.4.2 Chapter Review

FeCrNi and FeCo EML data sets for ground and space cover a wide range of undercoolings. Ground EML data had the highest levels of stirring although that stirring was driven by gravitational forces inherent in the levitation process. Shear rates required for RDM based model predictions can be produced through MHD modeling of internal fluid flows using known properties collected using ESL methods. Experimental delay times decrease with increasing undercooling and stirring in both alloys as shown in Figure 13 and Figure 14. For a given undercooling condition, experimental data fell within the range of delay time predictions where the upper limit is produced by low stirring and lower limit by high stirring conditions. To predict this range, high and low shear rates (indicative of high and low stirring respectively) were substituted into a Saturation Model for delay time. Model validity can be checked through an analysis of dimensionless parameters, while the validity of experimental delay times can be achieved through comparison of primary and secondary phase growth rates against values obtained from the literature.

Chapter 5. Discussion

5.1　Overview

Analysis of the figures presented in Chapter 4 allows for a better understanding of the solidification behavior of FeCo and FeCrNi alloys. Their differences are deeply rooted in internal melt shear present during solidification which is dependent on flow velocities that can be predicted by MHD modeling and are strongly influenced by externally applied stirring forces that can be generated from induced magnetic fields. The different reactions between alloys to applied convective forces can have a large impact on the material structure produced by comparatively long or short delay times. This chapter will first compare differences between the two RDM based convection models discussed in Chapter 3. This will be followed by a comparison of material differences between FeCo and FeCrNi identified through convection modeling.

5.2　Convection Model Comparison

One of the key purposes of this paper was to select an ideal model that can accurately predict delay time as a function of undercooling for a specified alloy composition across the ferrous alloy family. For each alloy mentioned above, two models were identified to closely match experimental data obtained from space and ground EML. Read-Shockley and optimized Saturation Models were both matched to the data sets for retained convective energy as a function of shear rate in FeCo and FeCrNi as shown in Figure 9. The Read-Shockley Model reaches a peak toward the median shear rate, then arches back down to fit the ground-based data at higher shear rates. This behavior is only possible if relaxation of recrystallization were to occur and the timeframe for these processes is orders of magnitude greater than observed for the metastable-stable transition. Thus, the prediction that a reduction in retained damage occurs at high convective conditions seems aphysical as the amount of retained convective energy stored in the system is known to scale with the defect energy i.e., the number of grain boundaries and with dislocation density, both of which should not decrease with increasing convection [13]. The wide spread of the space data in terms of retained convective energy for a given shear rate suggests that many of the higher data points might be outliers. This is more noticeable when observing the large concentration of ground-based data at a lower ΔG_C that is of equivalent value to the lower band of space-based data and aligned with the saturation curve.

In general, the Read-Shockley Model can be used to identify a local maximum ΔG_C at median shear rates whereas the Saturation Model more accurately predicts overall expected behavior across the entire range of $\dot{\gamma}$. Unlike FeCo, the FeCrNi data conforms to either model as seen in Figure 10. While the curve's peak was still fixed at the median shear rate, it was significantly shallower resulting in a near constant ΔG_C for values above $\dot{\gamma} = 250$ s^{-1}. The aphysical nature of the Read-Shockley curve was determinably less pronounced in FeCrNi. The saturation curve thus reflects a more intuitive data fit by providing a prediction for the limit of retained convective energy as shear rate is increased.

Normalizing the data set to examine Read-Shockley behavior from another perspective, we see in Figure 11 for FeCo the spread of data points is evenly distributed on either side of the Read-Shockley linear fit. Closer examination of the data points suggests that aside from two visible outliers, they follow a generally curved path which is more closely approximated by the saturation curve. Both Read-Shockley and saturation curves are also notably anchored about the ground data which thus represents critical data. Both ground and space data are required to characterize behavior. A normalized view of FeCrNi in Figure 12 again shows nearly identical curves, with the Saturation Model assuming a slightly more linear form which has an equal distribution of data points on both sides on the curve. The two models do not differ significantly and thus the FeCo data helps us to understand extremum behavior to a greater extent.

In all cases the Saturation Model has a marginally better fit than the Read-Shockley Model and provides a better physical perspective on retained damage energy behavior. As such it was selected for use as a predictor of delay times for the two alloys discussed. Assuming that experimental data follows the saturation curve closely, delay times computed from ΔG_C for FeCo are shown in Figure 13. The resulting plot displays predicted delay times for the range of shear conditions as calculated from experimental observation. For data points above 70 degrees undercooling, most of the data is either contained within or close to the bounds of the prediction. Doing the same for FeCrNi in Figure 14 suggests a similarly close prediction with a larger percentage of the points lying along the predicted trend lines, particularly at shallow undercooling. The closer the fit of the saturation curve to the data set, the better the predictor will be in determining delay time ranges.

5.3   Alloy Behavior Comparison

The secondary purpose for conducting the testing was to identify differences in convection driven solidification behavior between two unique alloys. Because convective damage models are largely dependent on internal shear rates as defined by MHD simulations, the root of any observed behavioral differences may lie in the thermophysical properties of density, viscosity, and electrical conductivity used to generate these simulations.

While density and viscosity are comparable between FeCrNi and FeCo, the slope of conductivity with temperature is notably inverted. This means that at deep undercoolings, the conductivity of FeCrNi is low and increases with increasing temperature, while conductivity for FeCo starts out significantly higher at deep undercoolings then is reduced with increasing temperature. The resultantly large difference in conductivities between FeCo and FeCrNi at deep undercoolings explains why the FeCrNi master equation for shear rate as a function of temperature (Equation 50) predicts a significant deviation from linearity at low temperature in Figure 8 when H_{CV} is set to maximum stirring conditions of 5.5 V, then becomes more linear as temperature is increased at constant voltage. Conversely, an equivalent plot of the simulation for FeCo in Figure 7, also at 5.5 V, shows shear rates increasing linearly across high and low temperatures alike. At higher temperatures (shallow undercooling) when conductivity converges between the two alloys, shear rates in Figure 7 and Figure 8 are likewise closer in magnitude and occur at similar velocities. But at deeper undercooling, FeCo has considerably larger shear rates than FeCrNi as exemplified by Figure 9 and Figure 10 where high stirring experimental data (Brown dots) are located further down the x-axis ($\dot{\gamma}$) for FeCo. Increased shear rates can be directly correlated to higher ΔG_C seen in Figure 9 4.50 V FeCo, compared to lower ΔG_C 4.50 V FeCrNi in Figure 10.

Differences in shear rates and convection between the two alloys can be further highlighted by novel optimized model parameters identified in Table 3 where FeCo was characterized by a steeper slope and maintained a greater y-intercept than FeCrNi using the Read-Shockley Model. The initial slope of the Saturation Model was similarly steeper for FeCo and reached a plateau at a 29.26% increased value for ΔG_C compared to FeCrNi. This suggests that by both model predictions, FeCo generally has more convective energy retained under equivalent undercooling conditions compared to FeCrNi.

5.4   Comments on Model Validity

Dimensionless analysis of delay times and driving force highlights the capability of Read Shockley and Saturation methods to accurately model convective damage in a manner that agrees with RDM theory for a slope of -4. The dependency of delay time on thermodynamic driving force is key to understanding material behavior as it provides a direct linkage between added energy due to convection and resulting material microstructure. Surprisingly, the Read Shockley generated slope between dimensionless parameters was within standard deviation of the slope for saturation in both alloys. Statistically, both models are accurate predictors of delay time for this reason. That being said, there is a clear benefit to the saturation approach in approximating a limit to convective free energy at higher shear rates.

Validity of experimental delays was reviewed through an investigation of growth velocities at the end of Chapter 4. Looking Back to Figure 20, there was a clear increase from experimental growth velocity of the secondary phase at 5.39 ± 0.39 as compared to the 2.40 ± 0.23 of experimental data. However, the predictions on lateral heat flux for the FeCo alloy come more in line with the positive deviation observed for other FeCo alloys [53]. Of particular importance in Figure 20 is the observation that growth of the stable austenite into the ferrite mushy zone is constant and independent of the original undercooling. This happens because following primary recalescence, and in the absence of significant partitioning, the mushy zone exists at the metastable liquidus no matter the observed original undercooling and thus subsequent growth occurs in all cases into liquid of the same temperature.

Chapter 6. Conclusions and Future Work

6.1 Utility of MHD Modeling Approach

One key utility of this work lies in the simplification of the procedure required to model shear rates for fluid flow conditions when only minimal information about the alloy such as the amount of stirring (inferred from H_{CV}) and temperature during solidification are known. Previous works have suggested that when processing solidification data, MHD surrogate models should be re-run for each set of test conditions present.

- It has been shown in Section 4.2 that a better approach to generating this shear data for experimental results would be to first run MHD simulations over a wide range of conditions for each test alloy. Then, by tracking dependencies of shear rate on key parameters such as temperature, flow velocity and H_{CV}, generate a set of alloy parameters for use with an alloy specific master equation like the one shown in Equation 50.

- Following this approach, the fluid flow conditions for novel experimental data can be quickly generated by simply entering the two experimental conditions of undercooling and H_{CV} into the master equation rather than waiting for a lengthy simulation to run its course.

- When a different alloy composition is tested, the parameters in the master equation need only be swapped out, allowing for significantly faster processing times.

6.2 Damage Model Comparison

A Saturation Model was contrasted against the Read-Shockley Model in terms of its ability to predict delay behavior in solidifying FeCo and FeCrNi.

- At low shear rates, the Saturation Model prediction closely matches the prediction of the Read Shockley Model.

- The Read Shockley Model exhibited prominent A-Physical Behavior when compared against the FeCo alloy dataset, with its parabolic trend in the ΔG_C vs. $\dot{\gamma}$ plot suggests a declining retained free energy function beginning at $\dot{\gamma} = 225$ s^{-1}. Shear rates declination is only possible if the material undergoes relaxation or recrystallization and the timeframe for this to occur seems too short. With these mechanisms becoming improbable, in this region as incrementally increased microstructural damage must yield progressively higher shear rates.

- The Saturation Model compensates for Read Shockley A-Physical Behavior at High Shear Rates, providing a closer prediction of delay behavior by gradually approaching but not exceeding a ΔG_C plateau of 9×10^7 J/m^3 for FeCo. This represents the maximum damage energy that may be stored in this material.

6.3   Alloy Comparison

Two Unique alloy compositions were examined using the Saturation Model. 3 unique differences in delay behavior between these alloys were noted.

- Differences in laminar flow velocity for FeCrNi and FeCo at equivalent undercooling conditions were observed through MHD modeling. The most significant of which being that FeCo has a 92% larger flow velocity than FeCrNi at deep undercoolings and high H_{CV} induced stirring.

- Available convective damage free energy in FeCo increased by 29.26% from FeCrNi. This effect is likely a byproduct of the increased flow velocity in FeCo.

- The Limit to available free energy in the Saturation Model for FeCrNi is $7 \pm 1.44 \times 10^7$ J/m^3 which is lower than the limit of FeCo at $9 \pm 2.28 \times 10^7$ J/m^3 but within the same order of magnitude. The standard deviations suggest that both materials yield a comparable fit.

6.4   Model Validation

To validate findings, a dimensionless quantity of delay time was produced for the data set using only high-speed video. A second dimensionless parameter of driving force was calculated the undercoolings recorded for the same data set using only pyrometry. These two parameters were plotted on a log log curve to verify their relationship. Dimensionless delay time is independent of dimensionless driving force.

- For FeCrNi the Saturation Model, a power curve fit to experimental data yielded an exponent of n which was close to the expected relation of -4 defined by RDM theory. Fitting the curve through the origin of the Log-Log plot yielded a linear slope of -3.98688 $\pm$ 0.031081 for the Saturation Model and -3.99769 $\pm$ 0.031139 for the Read Shockley Model. This is an indicator of a good fit in both high-speed camera generated data and delay data to the RDM theory for both models. Since the Read-Shockley model ties the Retained Damage Model to a physical effect related to dislocation damage, it is preferred for FeCrNi analysis.

- For FeCo the Saturation Model also yielded a power curve exponent n while the read Shockley model produced a power curve with slightly steeper slope. Fitting the curve through the origin of the Log-Log plot yielded a linear slope of -3.97953 $\pm$ 0.032794 for the Saturation Model and -4.00508 $\pm$ 0.033667 for the Read Shockley Model. Since both models return equally good agreement with theory, there is no strong evidence to support one or the other model and further work is required.

To validate experimental delay data, growth velocity was calculated for the primary and secondary phase in FeCo for the range of undercoolings present in the FeCo EML data set and compared to literature.

- Metastable phase velocities from our FeCo alloy are in close agreement with previous results across all tested undercoolings.

- Stable phase velocities from our FeCo alloy were significantly higher at 5.39 m/s compared to literature at 2.40 m/s. T-testing conducted on these two sample populations resulted in a clear rejection of the null hypothesis at 5% significance that the sample means are equivalent. This difference in the distribution of growth velocities for the two stable phase populations suggests that more data is needed to verify the correct growth velocity of FeCo.

6.5   Future Work

Ferrous alloys are key materials used in the manufacture of aerospace components and consumer goods. While thermophysical property measures attained using ESL methods have been widely distributed for the most used compositions of these alloys, not enough emphasis is being placed on fully understanding the mechanisms behind their solidification. Further development of RDM convective modeling such as the Saturation and Read Shockley approaches detailed in this book could allow for the microstructural evolution of ferrous alloys to not only be predicted but also influenced and enhanced through the targeted application of stirring to areas of interest during solidification. It is possible that combining predictions of solidification time as a function of internal convection could greatly enhance the process of mold design for casting processes. For example, fluid flows around the edges of a mold could be paired with estimates of solidification time leading to enhanced identification of miniscule areas of weakness in a cast product that could lead to failure of the part as a whole. A more achievable goal for the near term would be to apply the Saturation and Read Shockley Models to a wider experimental data set to both enhance current predictions, and expand the insights gained to other material compositions.

Chapter 7. References

[1]   D.M. Matson, X. Liu, J.E. Rodriguez, S. Jeon, and O. Shuleshova, "Retained free energy with enhanced nucleation during electrostatic levitation of undercooled Fe-Co alloys", *Crystals* **11** (2021) 1-10.

[2]   M. Volmer and A. Weber, "Nuclei formation in supersaturated systems", *Z. Phys. Chem.* **119** (1926) 227-301

[3]   R. Becker, "Nuclear formation in the separation of metallic mixed crystals" *Anal. Phys.* **32**[1-2] (1938) 128-140

[4]   J. C. Fisher, J. H. Hollomon, D. Turnbull, "Nucleation" *J. Appl. Phys.* **19** (1948) 775-784.

[5]   D. Turnbull, J.C. Fisher, "Rate of nucleation in condensed systems", *J. Chem. Phys.* **17** (1949) 71-73.

[6]   D. Turnbull, "Formation of crystal nuclei in liquid metals", *J. Appl. Phys.* **21** (1960) 1022-1028.

[7]   D. R. H. Jones and G. A. Chadwick, "An expression for the free energy of fusion in the homogeneous nucleation of solid from pure melts", *Phil. Mag.* **24** (1971) 995-998.

[8]   J. W. Christian, The theory of transformation in Metals and Alloys, (Oxford: Pergamon, 1975) 418.

[9]   C. V. Thompson and F. Spaepen, "On approximation of the free energy change on crystallization", *Acta Metall.* **27** (1979) 1855-1858.

[10]   D. E. Temkin and V. V. Shevelev, "On the theory of nucleation in two-component systems", *J. Cryst. Growth* **52** (1981) 104-110.

[11]   W. J. Boettinger and M. J. Aziz, "Theory for the trapping of disorder and solute in intermetallic phases by rapid solidification", **Acta Metall. 37**[12] (1989) 3379-3391.

[12]   M. Hillert, "Solute drag, solute trapping, and diffusional dissipation of Gibbs Energy" *Acta mater.* **47**[18] (1999) 4481-4505.

[13]   D.M. Matson, "Retained free energy as a driving force for phase transformation during rapid solidification of stainless steel alloys in microgravity", *npj Microgravity* **4:22** (2018) 1-6.

[14]   D.M. Matson, "Chapter 14 Influence of Convection on Phase Selection" H-J. Fecht, M. Mohr (eds.), Metallurgy in Space Recent, The Minerals, Metals & Materials Series.

[15]   R. Hermann, W. Löser, G. Lindenkreuz, A. Diefenbach, W. Zahnow, W. Dreier, T. Volkmann, D. Herlach, "Metastable phase formation in undercooled Fe-Co melts", *Mat. Sci. Engr.* **A375-377** (2004) 507-511.

[16]   R.W. Hyers, D.M. Matson, K.F. Kelton, and J.R. Rogers, "Convection in Containerless Processing", *Ann N.Y. Acad. Sci.* **1027** (2004) 474-494.

[17]   A. Seidel, W. Soellner, and C. Stenzel, 'EML - An electromagnetic levitator for the International Space Station', *Journal of Physics: Conference Series* **327** [1] (2011) 012057 1-14.

[18]   J.Brillo, and I. Egry, "Density Determination of Liquid Copper, Nickel, and Their Alloys", *International Journal of Thermophysics.* **24** (2003), 1155-1170

[19]   P.B. Coates "Multi Wavelength Pyrometry", *Meterologia,* **17** (1981), 103

[20]   J.E. Rodriguez, and D.M. Matson, "Lateral heat flux and remelting during growth into the mushy-zone", *Acta Mater.* (2017).

[21]   M. Barth, D. Holland-Moritz, and D.M. Herlach, "Dendrite growth velocity measurements in undercooled Ni and Ni-C melts in space", *The Minerals Metals & Materials Society Solidification* (1999).

[22]   C. Kreischer, and T. Volkmann, "Transformation kinetics of the metastable bcc phase during rapid solidification of undercooled Fe-Co alloy melts", *Materialia.* **20** (2021) 101211

[23]   W. T. Read and W. Shockley, "Dislocation models of crystal grain boundaries", *Phys. Rev.* **78** (1950) 275-289.

[24]   J. Feder, K.C. Russell, J. Lothe, G.M. Pound, "Homogeneous nucleation and growth of droplets in vapours", *Adv. Phys.* **15** (1966) 111–178.

[25]   A. Kantrowitz, "Nucleation in very rapid vapor expansions", *J. Chem. Phys.* **19** (1951) 1097-1100.

[26]   H. Wakeshima, "Time lag in the self-nucleation", *J. Chem. Phys.* **22** (1954) 1614-15.

[27]   F. C. Collins, "Time lag in spontaneous nucleation due to non-steady state effects", *Z. Elektrochem.* 59 (1955) 404-407.

[28]   S. Toschev and I. Gutzow, "Time lag in heterogeneous nucleation due to nonstationary effects*", Phys. Stat. Sol.* **21**[2] (1967) 683-691.

[29]   K. C. Russell, "Linked flux analysis of nucleation in condensed phases" *Acta Metall.* **16**, (1968) 761-769.

[30]   D. Kaschiev, "Solution of the non-steady state problem in nucleation kinetics", *Surface Science* **14** (1969) 209-220.

[31]   D. Kaschiev, "Nucleation at existing cluster size distributions", *Surface Science* **18** (1969) 389-397.

[32]   F. F. Abraham, "Multistate kinetics in nonsteady-state nucleation: a numerical solution", *J. Chem. Phys.* **51**[14] (1969) 1632-1638.

[33]   K. C. Russell, "Nucleation in solids: the induction and steady state effects", *Adv. Colloid and Interface Sci.* **13** (1980) 205-318.

[34]   K. F. Kelton, A. L. Greer, C. V. Thompson, "Transient nucleation in condensed systems", *J. Chem. Phys.* **79** (1983) 6261-6276.

[35]   G. Shao and P. Tsakiropoulos, "Prediction of phase selection in rapid solidification using time dependent nucleation theory", *Acta Metall. Mat.* **42**[9] (1994) 2937-2942.

[36]   C. Yang, F. Liu, G. Yang, Y. Chen, N. Liu, J. Li, and Y Zhou, "Non-equilibrium transformation in hypercooled Fe83B17 alloy", *Mat Sci. Engr. A* **458** (2007) 1-6.

[37]   D.M. Matson, "Influence of induced convection on transformation kinetics during rapid solidification of steel alloys: The Retained Damage Model", *JOM* **72** [11] (2020) 4109-4116.

[38]   J. Rodriguez, C. Kreischer, T. Volkmann and D.M. Matson, "Solidification velocity of Undercooled Fe-Co Alloys", *Acta Mater.* **122** (2017) 431-437.

[39]   J.E. Rodriguez and D.M. Matson, "Thermodynamic Modeling of the Solidification Path of Levitated Fe-Co Alloys", *CALPHAD* **49**[6] (2015), 87-100.

[40]   X. Xiao, J. Lee, R.W. Hyers, and D.M. Matson, "Numerical representations for flow velocity and shear rate inside electromagnetically levitated droplets in microgravity", *npj Microgravity* **5**:7 (2019), 1-7.

[41]   X. Xiao, R.W. Hyers and D.M. Matson, Surrogate model for convective flow inside electromagnetically levitated molten metal droplet using magnetohydrodynamic simulation and feature analysis", *Int. J. Heat and Mass Transfer* **136** (2019) pp. 532-542.

[42]   J. Lee, J.E. Rodriguez, R.H. Hyers, and D.M. Matson, "Measurement of Density of Fe-Co Alloys using Electrostatic Levitation", *Metallurgical and Materials Transactions B,* **46**[6] (2015) 2470-2475.

[43]   Y. Sato, K. Sugisawa, D. Aoki, and T. Yamamura, 'Viscosities of Fe-Ni, Fe-Co and Ni-Co binary melts', *Meas Sci Technol,* **16** [2] (2005) 363-371.

[44]   Y. Oichi Ono, T. Yagi, 'Electric Resistivity of Molten Fe-Ni and Fe-Co Alloys',
        Transactions of the Iron and Steel Institute of Japan, **12** [4] (1972) 314-316.

[45]   H. Tamaru, C. Koyama, H. Saruwatari, Y. Nakamura, T. Ishikawa, and T. Takada,
        "Status of the Electrostatic Levitation Furnace (ELF) in the ISS-KIBO", *Microgravity
        Science and Technology.* (2018), 30:643-651

[46]   A. Diefenbach, S. Schneider, and T. Volkmann, "Experiment Preparation and
        Performance for the Electromagnetic Levitator (EML) Onboard the International Space
        Station

[47]   I. Egry, A. Diefenbach, W. Dreier, J. Piller, "Containerless processing in space -
        Thermophsical property measurements using electromagnetic levitation", *Int. J.
        Thermophys.* **22** (2001) 569-578.

[48]   D.M. Matson, X. Xiao , J. Rodriguez and R.K. Wunderlich, "Preliminary Experiments
        Using Electromagnetic Levitation On the International Space Station", *International
        Journal of Microgravity Science and Application*, 33[2] (2016), 330206 1-11.

[49]   D. M. Matson, "The Measurement of Dendrite Tip Propagation Velocity During Growth
        into Undercooled Metallic Melts", in Solidification 1998, S. P. Marsh, J. A. Dantzig, R.
        Trivedi, W. Hofmeister, M. G. Chu, E. J. Lavernia, and J.-H. Chun, eds., TMS
        Warrendale PA, (1998), pp. 233-244.

[50]   D. Kirk, 'Saturation Curve Analysis and Quality Control', The Shot Peener magazine **20**
        [3] (2006) 24-30.

[51]   J. Zhang, H. Wang, W Kuang, Y. Zhang, S. Li, Y. Zhao, D. M. Herlach, "Rapid
        solidification of non-stoichiometric intermetallic compounds: modeling and experimental
        verification" *Acta Mater.* **148** (2018) 86-99.

[52]   T. Volkmann, W. Loser, and D. M. Herlach, "Nucleation and phase selection in
        undercooled Fe-Cr-Ni melts: Part I theoretical analysis of nucleation behavior", *Met.
        Trans.* **28A** (1997) 435-460.

[53]   J.E. Rodriguez and D.M. Matson, "Lateral Heat Flux during Remelt Growth into the
        Mushy-zone", *Acta Mater.* **129** (2017) 408-414.

[54]   J.E. Rodriguez, C. Kreischer, T. Volkmann, and D.M. Matson, "Solidification velocity of
        undercooled Fe-Co alloys", *Acta Mater.* (2016).

Appendices

Delay Time Code V4.8 – MATLAB R2021A

% This program opens a ISS-EML.avi sample video file and computes delay time between two phases
% Author: Brian Stanford

% The included script is an overhaul of the code originally designed by Willium Liu, and bug tested by Brian Stanford.
% Major Changes in this version include the assumption that the sample Sphere does not translate between frames
% To enact this change, the translationOfPOI function written by william has been removed and replaced with a series of nested for loops
% All cells have been replaced with matrices to allow for the easy printing of results to excel.
% This code has also been updated to include a skip feature which can reduce wasted time between attempts if a backup file has been generated in the past containing information.

% Main Script: Offers the user a chance to load information from a backup and skip any or all of the following parts
% Setup (Prompts 01, 02, 03): Load a video file, ask user for key frame information
% Part1 (Prompt 04): Identifies the outer edge of the sample, finds center, radius for each frame, then takes the average across all frames
% Part2 (Prompt 05): Allows the user to pick the origin of a phase
% Part3 (Prompt 06): Allows the user to Track points on the edge of a phase as they
% Part4 (calculations): Computes the 3-D Euclidean Distance between each x,y tracker point (from Part 06) and the x,y phase 1 or phase 2 centerpoint (from Part05) assuming a sphere with x,y center points and radius (from Part 04). Then computes the average Euclidean distance for each frame.
% Part5 (Plot): Plot the Euclidean distances (y) vs. frame Number (x) for phase 1 and 2, and determine a trendline. the distance between the x intercept of both trendlines is finaly computed in terms of frame numbers, then the frame number is converted to a time using the known video frame rate.
% Part6 (Adjustments): Experimental section to remove or adjust tracker locations,(Currently Incomplete)
% Part7 (Advanced Import): Planned feature to import and convert williams data file for comparison

%useful commands
%save('BackupFile', 'A', 'B') % Saves Matrices A and B to Backupfile.mat
%load('BackupFile', 'A', 'B') % loads matrices A and B from Backupfile.mat
%fprintf('message \n')   % prints a "message" to the command window then creates a new line with the \n command
%close all   % closes all open windows

%list of counters used in "for" loops, the prompt where they are located, and what variable the count
% (Name) - (Prompt)   (Variable)   (Description)
%counter01 - (prompt 1)   (Frame Number) counts up from 1 until it reaches the current frame number (note to self- find a way to create a matrix that lists the start frame in the first row)
%counter02 - (prompt 1)   (?????????????) unknown use... from williams code, maybe useful in later steps
%counter03 - (prompt 4)   (Frame Number) counts up starting with the start frame of phase 1 to the end frame of phase 2 - used to obtain circle center information for all frames; dfependant on phase start, end information
%counter04 - (prompt 4)   (Frame Number) same use as counter 01, starts at 1 and increments to save circle center information to a new row for each frame looked at (assumes row 1 = phase1 start frame)
%counter05 - (prompt 6)   (Tracker Number) counts up from 1 to the maximum number of trackers in phase 1 and 2. If phase 1 has more tracker than phase 2, then the number of trackers for phase 1 is used.... and vice versa. each time the counter loops, a new tracker variable will be saved, such that there is one matrix for each tracker containing information from that tracker from both phases
%counter05 old description: starts at 1 and increments up to the number of tracker locations specified by the maximum of prompt_06_A and prompt_06_B, also relies on counter01 to determine the number of video frames
%counter06 - (prompt 6)   (Frame Number) counts up from the video frame number when phase 1 starts to the video frame number when phase 1 ends. It is used in a for loop which opens up a new video frame as specified by this counter, then proceeds with a second loop to fill in tracker (06_A, 06_B)
%counter06_A - (Prompt 6)   (Tracker Number) counts up from 1 to the number of trackers in a for loop. This is paired with an if loop which makes use of this counter as the current tracker, and saves ginput data to that trackers matrix
%counter06_B - (Prompt 6)   (Frame Number) counts up from 1, recording the first frame in the phase as frame 1, then incrementing for each subsequent frame. data is saved to a row number specified by this counter in the tracker matrix
%counter07 - (prompt 6)   (Frame Number) counts up from the video frame number when phase 2 starts to the video frame number when phase 2 ends. It is used in a for loop which opens up a new video frame as specified by this counter, then proceeds with a second loop to fill in tracker (06_A, 06_B)
%counter07_A - (Prompt 6)   (Tracker Number) counts up from 1 to the number of trackers in a for loop. This is paired with an if loop which makes use of this counter as the current tracker, and saves ginput data to that trackers matrix
%counter07_B - (Prompt 6)   (Frame Number) counts up from 1, recording the first frame in the phase as frame 1, then incrementing for each subsequent frame. data is saved to a row number specified by this counter in the tracker matrix
%counter08 - (Calculations) (Tracker Number) Starts at 1 and increments up to the number of tracker locations specified by the maximum of prompt_06_A and prompt_06_B to run a calcualtion on each set of trackers

%Counter08_A.. (Calculations) (frame Number)   Each of these counters represents current frame number and counts up from either 1 or 2 until
it reaches the maximum frame in a given phase. Used in for loops, the counter can be referenced to read the specidied row of a matrix with each
row being a different frame numnber.
%counter09
%counter10
%counter11
%counter12
%counter13
%counter14
%counter15

%Brian Stanford 05/13/2022
%% Close figures, Clear Workspace, Clear Command Window

close all;
clear;
clc;

%% General Code - Data Import or Create New Project

%Prompt 01
cd 'C:\Users\Name\Desktop'; %change directory to users desktop (change 'user' based on computer being used)
prompt_01_FileName = 'Input the name of "ISS_EML.avi" file to be used in analysis: \n'; %creates the label for the dialogue box
fileName = inputdlg(prompt_01_FileName); %asks user to identify their video file on the desktop

% Initialize Variables
prompt_06_A = 0; % placeholder for a variable used in an else loop later on --> if data import fails the loop wont work.

%Prompt 02
prompt_02_DataImport = 'Would you like to import data? Y/N '; %creates the label for the dialogue box
dataImport = inputdlg(prompt_02_DataImport); %asks the user if they would like to import data respond with a y for yes or n for no
dataImportMat = cell2mat(dataImport);

    if dataImportMat == 'Y'   %if user wants to import, Ask for data import file name "data.m"
       prompt_02_A = 'Input the name of the "Data.mat" file saved to your desktop that you would like to import: '; %creates the label for the
dialogue box
       importFileName = inputdlg(prompt_02_A); %save users response (file name)
       importFileNameMat = cell2mat(importFileName);
    end

if dataImportMat == 'Y'   % The following prompts are only needed if user elected to import data from a file (user can select which data to
import and which data to rerun)

%Prompt 03
prompt_03_FrameImport = 'Import phase start, End frame Information? Y/N '; %creates the label for the dialogue box
frameImport = inputdlg(prompt_03_FrameImport);
frameImportMat = cell2mat(frameImport);

    if frameImportMat == 'Y' %if user wants to import frame details
       load(importFileNameMat, 'phase1Start', 'phase1End', 'phase2Start', 'phase2End', 'frameTotPhase1', 'frameTotPhase2', 'maxFrames') %load
frame details from their backup file selected in prompt 02
    end

%Prompt 04
prompt_04_CenterImport = 'Import circle center data file? Y/N '; %creates the label for the dialogue box
centerImport = inputdlg(prompt_04_CenterImport);
centerImportMat = cell2mat(centerImport);

    if centerImportMat == 'Y' %if user wants to import circle center details
       load(importFileNameMat, 'Rmin', 'Rmax', 'sThresh', 'segmentI', 'IThresh', 'centerI', 'radius', 'frameLength', 'centers', 'frameCenterAverage',
'frameradiusAverage', 'AdjustC1', 'AdjustC1Mat', 'newCenter', 'centersAverage') %load circle center details from their backup file selected in
prompt 02
    end

%Prompt 05
prompt_05_PhaseImport = 'Import phase center data file? Y/N '; %creates the label for the dialogue box
phaseImport = inputdlg(prompt_05_PhaseImport);

```matlab
phaseImportMat = cell2mat(phaseImport);

    if phaseImportMat == 'Y' %if user wants to import phase center details
        load(importFileNameMat, 'j', 'frame1', 'numEdgePointsP2', 'numEdgePointsP2Mat', 'centerP2Edge', 'sumXYP2', 'sumXP2', 'sumYP2',
'AvgXP2', 'AvgYP2', 'pointImage05', 'pointImage06', 'AdjustP2', 'AdjustP2Mat', 'satisfied02', 'AdjustP2_A', 'AdjustP2_A_Mat', 'pointImage07',
'pointImage08') %load phase2 center details from their backup file selected in prompt 02
        load(importFileNameMat, 'i'         , 'numEdgePointsP1', 'numEdgePointsP1Mat', 'centerP1Edge', 'sumXYP1', 'sumXP1', 'sumYP1',
'AvgXP1', 'AvgYP1', 'pointImage01', 'pointImage02', 'AdjustP1', 'AdjustP1Mat', 'satisfied01', 'AdjustP1_A', 'AdjustP1_A_Mat', 'pointImage03',
'pointImage04') %load phase1 center details from theri backup file
    end

%Prompt 06
prompt_06_TrackerImport = 'Import tracker point data file? Y/N '; %creates the label for the dialogue box
trackerImport = inputdlg(prompt_06_TrackerImport);
trackerImportMat = cell2mat(trackerImport);

    if trackerImportMat == 'Y' %if user wants to import tracker details
        load(importFileNameMat, 'figure', 'prompt_06_A', 'prompt_06_AMat', 'objectFrame2', 'prompt_06_B', 'prompt_06_BMat',
'maxNumTrackers', 'maxNumTrackersMat', 'currentFrameImageP1', 'baseFileName', 'fullFileName', 'baseFileName2', 'fullFileName2') %load
tracker details from their backup file selected in prompt 02
        load(importFileNameMat, 'tracker1Loc', 'tracker2Loc', 'tracker3Loc', 'tracker4Loc', 'tracker5Loc', 'tracker6Loc', 'tracker7Loc', 'tracker8Loc',
'tracker9Loc', 'tracker10Loc') %load matrices storing x,y  coordinates for each tracker as it moves from frame to frame
        load(importFileNameMat, 'point1A', 'point1B', 'point1C', 'point1D', 'point1E', 'point1F', 'point1G', 'point1H', 'point1I', 'point1J',  'point2A',
'point2B', 'point2C', 'point2D', 'point2E', 'point2F', 'point2G', 'point2H', 'point2I', 'point2J') %loads plotted points
    end


%Prompt 07 (extra options - only if successful data import of entire data file can the video be safely ignored)
    if dataImportMat == 'Y' && frameImportMat == 'Y' && centerImportMat == 'Y' && phaseImportMat == 'Y' && trackerImportMat == 'Y' %
only show this option if all data has been imported successfully
        prompt_07_SkipVid = 'Would you like to skip loading the video file? Y/N '; %gives user the option of skipping to result calculations using
their input data
        skipVid = inputdlg(prompt_07_SkipVid); %saves your response
        skipVidMat = cell2mat(skipVid);
        load(importFileNameMat, 'videoReader', 'videoPlayer', 'videoFrame', 'click') %'delXYZCenter', 'phaseInitialCoord', 'delXP1', 'delYP1',
'delXP2', 'delYP2', 'delZP1', 'delZP2', 'delT01', 'delT02', 'delT03', 'delT04', 'delT05', 'delT06', 'delT07', 'delT08', 'delT09', 'delT10',
'euclideanDistP1', 'euclideanDistP2', 'euclideanSumP1', 'euclideanSumP2' )
        counter01 = 0;
    else
        skipVidMat = 'N';

    end

%prompt 08 (extra options - only applies if tracker data file was imported, and user wants to overwrite a portion of that file regarding phase 2)
% add this to overwrite phase 2 center data when user reaches that section (only save phase 1 import)
    if trackerImportMat == 'Y'  % only show this option if all data has been imported successfully
        prompt_08_SkipPhase1 = 'Would you like to rerun your imported phase 2? Y/N '; %gives user the option of skipping to result calculations
using their input data
        skipPhase1 = inputdlg(prompt_08_SkipPhase1); %saves your response
        skipPhase1Mat = cell2mat(skipPhase1);
    else
        skipPhase1Mat = 'N';

    end

%prompt 09 (extra options - only applies if tracker data file was imported, and user wants to overwrite a portion of that file regarding phase 2)
% add this to overwrite phase 2 tracker data when user reaches that section (only save phase 1 import)
    if trackerImportMat == 'Y'  % only show this option if all data has been imported successfully
        prompt_09_SkipPhase1 = 'Would you like to re-run your imported phase 2 Trackers (skip phase 1 tracker selection)? Y/N '; %gives user the
option of skipping to result calculations using their input data
        skipPhase1T = inputdlg(prompt_09_SkipPhase1); %saves your response
        skipPhase1TMat = cell2mat(skipPhase1T);

        if skipPhase1TMat == 'Y'
            trackerImportMat = 'N'; %if user opts to skip phase 1 but rerun phase 2, change prompt 6 to allow for a point rerun  (phase 1 tracker
selection will only be run if skipPhase1TMat is 'N'. note that if prompt 6 was originally 'N' (data not imported), skipPhase1TMat will
automatically be set to 'N' allowing both phases to be run.
        end

    else
```

```matlab
        skipPhase1TMat = 'N';
    end


else
    frameImportMat = 'N';
    centerImportMat = 'N';
    phaseImportMat = 'N';
    trackerImportMat = 'N';
    skipVidMat = 'N';
    skipPhase1Mat = 'N';
    skipPhase1TMat = 'N';

end %uncomment this to place everything in a loop, causes errors with undefined variables


%% Alternative menu (change all answers to 1 for yes, 2 for no  (experimental code adds radial boxes instead of text entry to reduce user error)

%prompt_02_DataImport   = menu('Would you like to import data?','Yes','No');
%prompt_03_FrameImport  = menu('Enter phase start, end frame Information?','Yes','No');
%prompt_04_CenterImport = menu('Enter circle center data file?','Yes','No');
%prompt_05_PhaseImport  = menu('Enter phase center data file?','Yes','No');
%prompt_06_TrackerImport = menu('Enter tracker point data file?','Yes','No');
%prompt_07_SkipVid      = menu('Would you like to skip loading the video file?','Yes','No');
%prompt_08_SkipPhase1   = menu('Would you like to overwrite phase 2 data?','Yes','No');


%% Load video file and check for total number of frames (prompt 01)

if skipVidMat == 'Y' %check if video was skipped
    fprintf('You have skipped the video! \n')

else %continue loading file if video was not skipped

 % code for video reading provided by William

   % initialize variables
      videoReader = VideoReader(fileName{1,1}); %reads the video file selected
      videoPlayer = vision.VideoPlayer;       %opens video player
      videoFrame = struct('cdata', zeros(videoReader.Height,videoReader.Width,3, 'uint8'));
      counter01 = 1; %current frame number counts up until it reaches the total number of frames in the video
      counter02 = 0;
      click = 0; %detects a user click when scrolling through frames

   % Count the number of frames in the video file
   while hasFrame(videoReader)
      videoFrame(counter01).cdata = readFrame(videoReader);
      videoPlayer(videoFrame(counter01).cdata)
      counter01 = 0;
      click = waitforbuttonpress;  % detect mouse click
      if click == 0  % on detect make y=1
         counter02 = 1;
         counter01 = counter01 + 1; %increment counter
      end
   end

   %close out of video viewer
   close all;

end

%% Prompt 01 results

   fprintf('you have successfully Loaded a Video with the following number of frames! \n')
   disp(counter01) %number of frames

   fprintf('Prompt 01 Complete! \n')


%% Notify User of Data Import (prompt 02)
```

```matlab
if dataImportMat == 'Y'
    fprintf('you have successfully Selected a file for data transfer! \n')
else
    fprintf('you have decided not to import data! \n')
end

    fprintf('Prompt 02 Complete! \n')


%% Enter Start and End Frames For Each Phase (prompt 03) --> The results of this section will be used in section 6

if frameImportMat == 'Y'                          %check if user imported data from general code section
    fprintf('you have successfully imported frame data! \n') %notify user of successful import, \n skips to new line
    %add line to print data variables
else                                    %else begins Frame Number code up until the end function
    %code to allow user to enter frame info goes here (could turn everything inside the else loop into a called function for use in other scripts)

    %ask user which frames to select
        prompt_03_A = 'Input the frame# when the first phase starts: \n';
        prompt_03_B = 'Input the frame# when the fist phase ends: \n';
        prompt_03_C = 'Input the frame# when the second phase starts: \n';
        prompt_03_D = 'Input the frame# when the second phase ends: \n';

    %save critical frame information %might need to convert cell2 mat and update everywhere
        phase1Start = input(prompt_03_A);
        phase1End   = input(prompt_03_B);
        phase2Start = input(prompt_03_C);
        phase2End   = input(prompt_03_D);

    %determine the Total number of frames in phase 1
        frameTotPhase1 = (phase1End - phase1Start + 1); %eg.  start frame 5, end frame 10 --> (frame10-frame5) = 5rows   so starting with
frame5...    frame5 = row1,  frame6 = row2,  frame7 = row3,  frame8 = row4,  frame9 = row5 .... but frame 10 is the last frame so need one extra
row... frame10 = row6 ..... therefore the actual calculation is    #rows required = {(frame end - frame start) + 1row}
    %determine the Total number of frames in phase 2
        frameTotPhase2 = (phase2End - phase2Start + 1);
    %determine the maximum number of rows (whichever has more frames phase 1 or phase 2)
        if frameTotPhase1 > frameTotPhase2 %if more frames in phase 1 than in phase 2
            maxFrames = frameTotPhase1; % then choose number of frames in phase 1 to be the max number of frames
        elseif frameTotPhase1 == frameTotPhase2
            maxFrames = frameTotPhase1;
        elseif frameTotPhase1 < frameTotPhase2
            maxFrames = frameTotPhase2;
        end

end

%% Prompt 03 results

    fprintf('you have successfully Generated frame data! \n')
        fprintf('Phase 1 Start \n')
            disp(phase1Start)
        fprintf('Phase 1 End \n')
            disp(phase1End)
        fprintf('Total Frames in Phase 1 \n')
            disp(frameTotPhase1)
        fprintf('Phase 2 Start \n')
            disp(phase2Start)
        fprintf('Phase 2 End \n')
            disp(phase2End)
        fprintf('Total Frames in Phase 2 \n')
            disp(frameTotPhase2)

        fprintf('Max Number of Frames in Phase 1 and 2 \n')
            disp(maxFrames)

    fprintf('Prompt 03 Complete! \n')


%% Enter Circle Center Information (prompt 04)
```

```matlab
if centerImportMat == 'Y'
    fprintf('you have successfully imported circle center data! \n')
    %add line to print data variables
else
    %code opens video file and picks out circle centers for each frame

    %Initialize Variables
        Rmin = 75;
        Rmax = 900;
        sThresh = 0.1;
        segmentI = 0;
        IThresh = 0;
        centerI = zeros(10, 2); %matrix with 10 rows and 2 columns (10 being arbitrarily large enough to hold all identified circles in a single video
frame and 2 representing x,y.
        radius = zeros(10,1); %same as centerI but with onlu 1 column representing radius

        % New Prompt Version 4.8
        promptCutVid = 'Would you like to only analyze the first 10 frames of phase 1 for sample center? Y/N '; %User can now choose whether to
average the center points for all frames in the video (could be up to 200 frames in helium) which must each be checked manually, or only obtain
sample centers for the first 10 frames as an approximate
            cutVid = inputdlg(promptCutVid); %store users response in a cell(Y or N)
            cutVidMat = cell2mat(cutVid); %converts the user response to a matrix so it can be used in if/then statements
        if cutVidMat == 'Y' %new to version 4.8,
            frameLength = (9 + 1); % 9 frames plus the first frame of phase 1 = 10 frames. this will set the length of the centers matrix to 10 rows
            trimmedVidEnd = phase1Start + 9 ; %counter 03 will now cycle from phase 1 start through to 9 frames after that frame (calculating 10
total)
        else %the else here used to be the only possible response prior to version 4.8
            frameLength = (phase2End - phase1Start + 1); %the length of the centers matrix will be set to the total numbers of frames from the start
of the first phase to the end of the second phase
            trimmedVidEnd = phase2End  ; %counter 03 will now cycle from phase 1 start through to phase 2 end when calculating center
coordinates
        end

        centers = zeros(frameLength, 3); %contains a row for each frame, and columns for x, y, radius center coordinates
        %counter03 = 0; % dont need to initialize variables used as a for loop condition
        counter04 = 1;  % used to save the final matrix of center coordinates


    for counter03 = phase1Start : trimmedVidEnd %used to be "phase1Start : phase2End" but for helium cycles where there are hundreds of
frames, time can be saved by running only the first X frames after the start of the first phase, in this case "Phase1Start : Phase1Start + 10"

        %isolate one frame of video
        segmentI = read(videoReader, counter03); %read current frame from counter 03 which is incremented from the start of phase 1 to end of
phase 2
        figure; %calculate figure
        imshow(segmentI); %display figure
        fprintf('Isolated Frame \n'); %let user know frame isolation was successful.
        %segment frame
        Ithresh = im2bw(segmentI, sThresh); %convert image to black and white
        figure;
        imshow(Ithresh);

        %find centroid using imfindcircles
        [centerI, radius] = imfindcircles(Ithresh, [Rmin Rmax], 'Sensitivity', 0.95);  %circle center information will be saved with one row in the
centerI and radius matrices for each circle identified.
        viscircles(centerI,radius); %display circles
        hold on;
        plot(centerI(:,1),centerI(:,2),'yx','LineWidth',2); %add text to the plot for circle center coordinates and radius

        %compute average for the frame
        %frameCenterAverage = mean(centerI); %this code doesnt work because it includes cells with zeros in the average... fix below%calculates
the average of all values in each column (column 1= x coord) (column2 = y coord) across all rows which represent detected circles. The result
will be a single row matrix with one x and one y coordinate
        %frameradiusAverage = mean(radius); %this line is also broken ... same reason as above, fix below
        frameCenterAverage = sum(centerI,1) ./ sum(centerI~=0,1); %need to check to make sure this works --> should include zeros in the average
        frameradiusAverage = sum(radius,1) ./ sum(radius~=0,1); %need to check to make sure this works

        %save averaged center and radius data to a new matrix with a row for each frame
```

```matlab
        centers (counter04,1) = frameCenterAverage (1,1);
        centers (counter04,2) = frameCenterAverage (1,2);
        centers (counter04,3) = frameradiusAverage (1,1);

        %Plot averaged center result
        viscircles(centerI,radius); %display circles
        plot(centers(counter04,1),centers(counter04,2),'yx','LineWidth',2);

        %make manual adjustment to the average?
        prompt_04_A = 'Adjust the center point? Y/N';  %prompt user to modify circle center manually
        AdjustC1 = inputdlg(prompt_04_A);
        AdjustC1Mat = cell2mat(AdjustC1);

        %use ginput to overwrite computer generated results
        if AdjustC1Mat == 'Y'
            %isolate one frame of video  %already loaded the figure so this block is redundant --> ignore
            %segmentI = read(videoReader, counter03); %read current frame from counter 03 which is incremented from the start of phase 1 to end
of phase 2
            %figure %calculate figure
            %imshow(segmentI);

            %Centers Manual update
            zoom on; %turn on the zoom feature
            pause %allow for zooming in until button is pressed
            [centers(counter04, 1), centers(counter04, 2)] = ginput(1); %allow the user to click on a point. Save that point to centers matrix in the
same spot where the original points were
            centers(counter04, 1) = round (centers(counter04, 1)); %update the x coordinate of the tracker to a rounded value
            centers(counter04, 2) = round (centers(counter04, 2)); %update the y coordinate of the tracker to a rounded value
            zoom out; %turn off the zoom feature, return to original image
            newCenter = insertMarker(segmentI, [centers(counter04, 1), centers(counter04, 2)],'+', 'Color', 'green'); %plot the now rounded
coordinates
            imshow(newCenter)
        end  %end manual center adjustment for the current frame

        hold off;
        counter04 = counter04 + 1; %increment counter 04 by 1 so that the next frame saves to a new row in the centers matrix created above

    end  %finished running each of the frames specified, at least one center found for each.


        %Final Averaged result with edits
        centersAverage = mean(centers); %Average together phase data
        centersAverage = round(centersAverage);


end

%% Prompt 04 Results

    % final results of prompt 4 is a single averaged center coordinate and sample radius for the applicable segment of video
    fprintf('Average Center Information (x, y, radius) for all frames in Phase 1 and 2! \n')
        disp(centersAverage)

    fprintf('Prompt 04 Complete! \n')

    %added to close all windows
    close all


%% Enter Phase Center Information (prompt 05)

if phaseImportMat == 'Y' %skip both phases if user elected to do so
    fprintf('you have successfully imported phase center data! \n')

else %continue to entry of corner points and phase center calculation for both phases

    if skipPhase1Mat == 'Y' %Secondary if statement skips phase 1 and only modifies phase 2
        fprintf('you have successfully skipped phase 1! \n') %notify user of successful import, \n skips to new line
```

```matlab
%Phase 2 - plot corner points to find the centroid of the phase
    %code identifies a center of each phase then allows the user to correct that selection
    %load video of phase 1 start + 1
        j = (phase2Start + 1);
        frame2 = read(videoReader, j);
        imshow(frame2);
    %select the number of corner points in phase 1
        prompt_05_D = 'How many corner Points do you see?';  %prompt user for number of corner points and generate a matrix to fill with
ginput coordinates
        numEdgePointsP2 = inputdlg(prompt_05_D);
        numEdgePointsP2Mat = cell2mat(numEdgePointsP2); numEdgePointsP2Mat = str2double(numEdgePointsP2Mat);
    %generate a empty matrix with the same number of rows as corner points selected (one row for each ginput coordinate pair) and (one
ginput coordinate pair for each corner point)
        centerP2Edge = zeros(numEdgePointsP2Mat, 2);
    %ginput for cooridinates and save to the centerP1Edge matrix
        zoom on; %added this line in to allow zooming in on small images
        pause %allow for zooming in until button is pressed
        [centerP2Edge(:,1),centerP2Edge(:,2)] = ginput(numEdgePointsP1Mat); %colon represents the row changes based on current
numEdgePointsP1Mat which was taken from user prompt to identify the number of edge points to track.
        zoom out; %cancel zoom once all points have been selected to save the full image
    %round coordinate values  into closest whole number (AKA Pixel)
        centerP2Edge(:, 1) = round(centerP2Edge(:, 1));
        centerP2Edge(:, 2) = round(centerP2Edge(:, 2));
    %find centroid of triangle
        sumXYP2 = sum(centerP2Edge); %check, avgXYP1 should form a row matrix with 3 columns, each containing a sum of values in said
column of centerP1Edge
        sumXP2 = sumXYP2(1,1); %extracts the x averaged value from the matrix
        sumYP2 = sumXYP2(1,2); %extracts the y averaged value from the matrix
        AvgXP2 = sumXP2 / numEdgePointsP2Mat; %takes average by dividing sum over number of points
        AvgYP2 = sumYP2 / numEdgePointsP2Mat; %takes average by dividing sum over number of points
    %plot points from ginput
        pointImage05 = insertMarker(frame2, [centerP2Edge(:,1), centerP2Edge(:,2)],'+', 'Color', 'green'); %creates coordinates to plot
        pointImage06 = insertMarker(frame2, [AvgXP2, AvgYP2],'+', 'Color', 'blue'); %creates coordinates to plot
        hold on %prepares to retain the images
        imshow(pointImage05); %plot edges
        imshow(pointImage06); %plot center
        hold off;
    %make Adjustments?
        prompt_05_E = 'Adjust X,Y? Y/N'; %prompt user to modify phase center manually
        AdjustP2 = inputdlg(prompt_05_E);
        AdjustP2Mat = cell2mat(AdjustP2);
    %loop for correction begins
    if AdjustP2Mat == 'Y' %if user responds yes, the tracker selection process is redone
        fprintf('you are unsatisfied with the computers Phase1 center selection! \n')
        satisfied02 = 'N' ; %default status of user is unsatisfied with result of averaged center
        while satisfied02 == 'N' %keep repeating prompt until user selects "N"
            prompt_05_F = 'Still unsatisfied? Select a new point! Y/N';  %prompt user to modify phase center
            AdjustP2_A = inputdlg(prompt_05_F);
            AdjustP2_A_Mat = cell2mat(AdjustP2_A);
          if AdjustP2_A_Mat == 'Y'
            %ginput to directly set a center point
                zoom on; %added this line in to allow zooming in on small images
                pause %allow for zooming in until button is pressed
                [AvgXP2,AvgYP2] = ginput(numEdgePointsP2Mat); %colon represents the row changes based on current
numEdgePointsP1Mat which was taken from user prompt to identify the number of edge points to track.
                zoom out; %cancel zoom once all points have been selected to save the full image
            %round coordinate values  into closest whole number (AKA Pixel)
                AvgXP2 = round(AvgXP2);
                AvgYP2 = round(AvgYP2);
            %plot points from ginput
                pointImage07 = insertMarker(frame2, [centerP2Edge(:,1), centerP2Edge(:,2)],'+', 'Color', 'green'); %creates coordinates to plot
                pointImage08 = insertMarker(frame2, [AvgXP2, AvgYP2],'+', 'Color', 'red'); %creates coordinates to plot
                hold on %prepares to retain the images
                imshow(pointImage07); %plot edges
                imshow(pointImage08); %plot center
                hold off;
          else %if user selects "N" to the querry of whether they are still unsatisfied (or enters any key other than capital Y) the satisfaction
status will change to "Y"
                satisfied02 = 'Y'; %exit while loop once user declares "N" to
```

```matlab
            end %end of if statement, continue to repeat if while is still true
        end %end of while loop
    end % end of original if statement for adjusting Phase 2


else %else statement requires points for both phase 1 and 2 to be entered
    fprintf('Begin Circle Center Location for Phases 1 and 2! \n') %notify user of successful import, \n skips to new line


    %Begin Phase 1 Point Tracking


    %Phase 1 - plot corner points to find the centroid of the phase
        %code identifies a center of each phase then allows the user to correct that selection
        %load video of phase 1 start + 1
        i = (phase1Start + 1);
        frame1 = read(videoReader, i);
        imshow(frame1);
        %select the number of corner points in phase 1
        prompt_05_A = 'How many corner Points do you see?'; %prompt user for number of corner points and generate a matrix to fill with
ginput coordinates
        numEdgePointsP1 = inputdlg(prompt_05_A);
        numEdgePointsP1Mat = numEdgePointsP1; numEdgePointsP1Mat = str2double(numEdgePointsP1Mat);
        %generate a empty matrix with the same number of rows as corner points selected (one row for each ginput coordinate pair) and (one
ginput coordinate pair for each corner point)
        centerP1Edge = zeros(numEdgePointsP1Mat, 2);
        %ginput for cooridinates and save to the centerP1Edge matrix
        zoom on; %added this line in to allow zooming in on small images
        pause %allow for zooming in until button is pressed
        [centerP1Edge(:,1),centerP1Edge(:,2)] = ginput(numEdgePointsP1Mat); %colon represents the row changes based on current
numEdgePointsP1Mat which was taken from user prompt to identify the number of edge points to track.
        zoom out; %cancel zoom once all points have been selected to save the full image
        %round coordinate values  into closest whole number (AKA Pixel)
        centerP1Edge(:, 1) = round(centerP1Edge(:, 1));
        centerP1Edge(:, 2) = round(centerP1Edge(:, 2));
        %find centroid of triangle
        sumXYP1 = sum(centerP1Edge); %check, avgXYP1 should form a row matrix with 3 columns, each containing a sum of values in said
column of centerP1Edge
        sumXP1 = sumXYP1(1,1); %extracts the x averaged value from the matrix
        sumYP1 = sumXYP1(1,2); %extracts the y averaged value from the matrix
        AvgXP1 = sumXP1 / numEdgePointsP1Mat; %takes average by dividing sum over number of points
        AvgYP1 = sumYP1 / numEdgePointsP1Mat; %takes average by dividing sum over number of points
        %plot points from ginput
        pointImage01 = insertMarker(frame1, [centerP1Edge(:,1), centerP1Edge(:,2)],'+', 'Color', 'green'); %creates coordinates to plot
        pointImage02 = insertMarker(frame1, [AvgXP1, AvgYP1],'+', 'Color', 'blue'); %creates coordinates to plot
        hold on %prepares to retain the images
        imshow(pointImage01); %plot edges
        hold on
        imshow(pointImage02); %plot center
        hold off;
        %make Adjustments?
        prompt_05_B = 'Adjust X,Y? Y/N'; %prompt user to modify phase center manually
        AdjustP1 = inputdlg(prompt_05_B);
        AdjustP1Mat = cell2mat(AdjustP1);
        %loop for correction begins
        if AdjustP1Mat == 'Y' %if user responds yes, the tracker selection process is redone
            fprintf('you are unsatisfied with the computers Phase1 center selection! \n')
            satisfied01 = 'N' ; %default status of user is unsatisfied with result of averaged center
            while satisfied01 == 'N' %keep repating prompt until user selects "N"
                prompt_05_C = 'Still unsatisfied? Select a new point! Y/N'; %prompt user to modify phase center
                AdjustP1_A = inputdlg(prompt_05_C);
                AdjustP1_A_Mat = cell2mat(AdjustP1_A);
                if AdjustP1_A_Mat == 'Y'
                    %ginput to directly set a center point
                    zoom on; %added this line in to allow zooming in on small images
```

```matlab
        pause %allow for zooming in until button is pressed
        [AvgXP1,AvgYP1] = ginput(numEdgePointsP1Mat); %colon represents the row changes based on current
numEdgePointsP1Mat which was taken from user prompt to identify the number of edge points to track.
        zoom out; %cancel zoom once all points have been selected to save the full image
    %round coordinate values  into closest whole number (AKA Pixel)
        AvgXP1 = round(AvgXP1);
        AvgYP1 = round(AvgYP1);
    %plot points from ginput
        pointImage03 = insertMarker(frame1, [centerP1Edge(:,1), centerP1Edge(:,2)],'+', 'Color', 'green'); %creates coordinates to plot
        pointImage04 = insertMarker(frame1, [AvgXP1, AvgYP1],'+', 'Color', 'red'); %creates coordinates to plot
        hold on %prepares to retain the images
        imshow(pointImage03); %plot edges
        hold on
        imshow(pointImage04); %plot center
        hold off;
    else %if user selects "N" to the querry of whether they are still unsatisfied (or enters any key other than capital Y) the satisfaction
status will change to "Y"
        satisfied01 = 'Y'; %exit while loop once user declares "N" to
    end %end of if statement, continue to repeat if while is still true
    end %end of while loop
end % end of original if statement for adjusting Phase 1




%continue on to phase 2 point tracking




%Phase 2 - plot corner points to find the centroid of the phase
    %code identifies a center of each phase then allows the user to correct that selection
    %load video of phase 1 start + 1
    j = (phase2Start + 1);
    frame2 = read(videoReader, j);
    imshow(frame2);
%select the number of corner points in phase 1
    prompt_05_D = 'How many corner Points do you see?';  %prompt user for number of corner points and generate a matrix to fill with
ginput coordinates
    numEdgePointsP2 = inputdlg(prompt_05_D);
    numEdgePointsP2Mat = cell2mat(numEdgePointsP2); numEdgePointsP2Mat = str2double(numEdgePointsP2Mat);
    %generate a empty matrix with the same number of rows as corner points selected (one row for each ginput coordinate pair) and (one
ginput coordinate pair for each corner point)
    centerP2Edge = zeros(numEdgePointsP2Mat, 2);
    %ginput for cooridinates and save to the centerP1Edge matrix
    zoom on; %added this line in to allow zooming in on small images
    pause %allow for zooming in until button is pressed
    [centerP2Edge(:,1),centerP2Edge(:,2)] = ginput(numEdgePointsP2Mat); %colon represents the row changes based on current
numEdgePointsP1Mat which was taken from user prompt to identify the number of edge points to track.
    zoom out; %cancel zoom once all points have been selected to save the full image
    %round coordinate values  into closest whole number (AKA Pixel)
    centerP2Edge(:, 1) = round(centerP2Edge(:, 1));
    centerP2Edge(:, 2) = round(centerP2Edge(:, 2));
    %find centroid of triangle
    sumXYP2 = sum(centerP2Edge); %check, avgXYP1 should form a row matrix with 3 columns, each containing a sum of values in said
column of centerP1Edge
    sumXP2 = sumXYP2(1,1); %extracts the x averaged value from the matrix
    sumYP2 = sumXYP2(1,2); %extracts the y averaged value from the matrix
    AvgXP2 = sumXP2 / numEdgePointsP2Mat; %takes average by dividing sum over number of points
    AvgYP2 = sumYP2 / numEdgePointsP2Mat; %takes average by dividing sum over number of points
    %plot points from ginput
        pointImage05 = insertMarker(frame2, [centerP2Edge(:,1), centerP2Edge(:,2)],'+', 'Color', 'green'); %creates coordinates to plot
        pointImage06 = insertMarker(frame2, [AvgXP2, AvgYP2],'+', 'Color', 'blue'); %creates coordinates to plot
        hold on %prepares to retain the images
        imshow(pointImage05); %plot edges
        hold on
        imshow(pointImage06); %plot center
        hold off;
    %make Adjustments?
        prompt_05_E = 'Adjust X,Y? Y/N';  %prompt user to modify phase center manually
```

```matlab
            AdjustP2 = inputdlg(prompt_05_E);
            AdjustP2Mat = cell2mat(AdjustP2);
        %loop for correction begins
        if AdjustP2Mat == 'Y' %if user responds yes, the tracker selection process is redone
            fprintf('you are unsatisfied with the computers Phase1 center selection! \n')
            satisfied02 = 'N' ; %default status of user is unsatisfied with result of averaged center
          while satisfied02 == 'N' %keep repating prompt until user selects "N"
            prompt_05_F = 'Still unsatisfied? Select a new point! Y/N'; %prompt user to modify phase center
            AdjustP2_A = inputdlg(prompt_05_F);
            AdjustP2_A_Mat = cell2mat(AdjustP2_A);
          if AdjustP2_A_Mat == 'Y'
            %ginput to directly set a center point
                zoom on; %added this line in to allow zooming in on small images
                pause %allow for zooming in until button is pressed
                [AvgXP2,AvgYP2] = ginput(numEdgePointsP2Mat); %colon represents the row changes based on current
numEdgePointsP2Mat which was taken from user prompt to identify the number of edge points to track.
                zoom out; %cancel zoom once all points have been selected to save the full image
            %round coordinate values  into closest whole number (AKA Pixel)
                AvgXP2 = round(AvgXP2);
                AvgYP2 = round(AvgYP2);
            %plot points from ginput
                pointImage07 = insertMarker(frame2, [centerP2Edge(:,1), centerP2Edge(:,2)],'+', 'Color', 'green'); %creates coordinates to plot
                pointImage08 = insertMarker(frame2, [AvgXP2, AvgYP2],'+', 'Color', 'red'); %creates coordinates to plot
                hold on %prepares to retain the images
                imshow(pointImage07); %plot edges
                hold on
                imshow(pointImage08); %plot center
                hold off,
          else %if user selects "N" to the querry of whether they are still unsatisfied (or enters any key other than capital Y) the satisfaction
status will change to "Y"
                satisfied02 = 'Y'; %exit while loop once user declares "N" to
          end %end of if statement, continue to repeat if while is still true
          end %end of while loop
        end % end of original if statement for adjusting Phase 2




    %completed Phase 2




    end %end if loop to skip phase 1 and else statement for when both phases 1 and 2 are entered
end %end if loop which allowed the user to skip both phases

%% Prompt 05 Results

    %load phase 1 image
    %load phase 2 image

    fprintf('Phase 1 Center X coordinate found! \n')
        disp(AvgXP1)
    fprintf('Phase 1 Center y coordinate found! \n')
        disp(AvgYP1)
    fprintf('Phase 2 Center x coordinate found! \n')
        disp(AvgXP2)
    fprintf('Phase 2 Center y coordinate found! \n')
        disp(AvgYP2)

    fprintf('Prompt 05 Complete! \n')


%% Enter Tracker Information (Prompt 06)

if trackerImportMat == 'Y'
    fprintf('you have successfully imported tracker data! \n')
    %add line to print data variables
```

```matlab
else
    %code to allow the user to slect points along the edge of the phase in each frame of the video with ginput

    %load and display a video frame of phase1 to determine number of trackers
        objectFrame1 = read(videoReader, phase1Start+2); %loads 2 frames after the phase 1 start frame
        objectRegion1 = [50, 70, 175, 175]; %loads a image window to display the desired frame
        figure; %saves as a temporary figure
        imshow(objectFrame1);
    %Select the number of trackers (up to 10) for phase1
        prompt_06_A = menu('How many trackers do you want to set for Phase 1?','1','2','3','4','5','6','7','8','9','10');
        prompt_06_AMat = prompt_06_A;

    %load and display a video frame of phase 2 to determine number of trackers
        objectFrame2 = read(videoReader, phase2Start+2); %loads 2 frames after the phase 1 start frame
        objectRegion2 = [50, 70, 175, 175]; %loads a image window to display the desired frame
        figure; %saves as a temporary figure
        imshow(objectFrame2);
    %Select the number of trackers (up to 10) for phase2
        prompt_06_B = menu('How many trackers do you want to set for Phase 2?','1','2','3','4','5','6','7','8','9','10');
        prompt_06_BMat = prompt_06_B;


    close all; %close the imageviewer frame

    %need to find a way to only perform the calculation if different numbers of trackers for phase 1 and phase 2 (otherwise if more trackers in
    phase 2, not enough matrices will be saved to record responses.
    %tentative solution, use only the maximum number of trackers from phase 1 and 2
        if prompt_06_AMat > prompt_06_BMat     %if phase1 (prompt 6A) has more trackers than phase 2 (prompt 6B), then calculations will
    continue through to the number of trackers in pahse 1
            maxNumTrackers = prompt_06_AMat; % define the max number of trackers for use in counter 7 loop. Note, calculations will be
    performed on empty cells for phase 2 which may or may not be a future problem.
            maxNumTrackersMat = maxNumTrackers;
        elseif prompt_06_AMat == prompt_06_BMat
            maxNumTrackers = prompt_06_AMat;
            maxNumTrackersMat = maxNumTrackers;
        elseif prompt_06_AMat < prompt_06_BMat
            maxNumTrackers = prompt_06_BMat;
            maxNumTrackersMat = maxNumTrackers;
        end

    %the following loops select how many tracker matrices are required
    %then create a matrix of x, y coordinates for each tracker
    %if prompt_06_A == 1 % if one tracker selected from the menu above, then only one tracker matrix required
    %sets up a matrix with a row for each video frame,  and  columns for  xphase1,yphase1,xphase2, yphase2 of a specific tracker.

    %Define the number of required tracker matrices
    for counter05 = 1 : maxNumTrackersMat %this loop will create tracker matrices for each tracker with the number of rows specified by the
    number of frames
        if counter05 == 1
            tracker1Loc= zeros(maxFrames, 4); % 4 columns (phase 1 x, phase 1 y, phase 2 x, phase 2 y) %note to self, maxFrames used to just be
    counter01 which made the matrix way too big (one cell for each frame) only really need one cell for each frame in the phase in this case... cell 1
    represents framestart #x
        elseif counter05 == 2
            tracker2Loc= zeros(maxFrames, 4); % create a matrix for the second tracker of both phase 1 and phase 2
        elseif counter05 == 3
            tracker3Loc= zeros(maxFrames, 4);
        elseif counter05 == 4
            tracker4Loc= zeros(maxFrames, 4);
        elseif counter05 == 5
            tracker5Loc= zeros(maxFrames, 4);
        elseif counter05 == 6
            tracker6Loc= zeros(maxFrames, 4);
        elseif counter05 == 7
            tracker7Loc= zeros(maxFrames, 4);
        elseif counter05 == 8
            tracker8Loc= zeros(maxFrames, 4);
        elseif counter05 == 9
            tracker9Loc= zeros(maxFrames, 4);
        elseif counter05 == 10
            tracker10Loc= zeros(maxFrames, 4);
```

```matlab
        end % end the if statement
    end % end the for loop



        fprintf('Begin Phase 1 Tracker Selection! \n') %notify user of begin ginput, \n skips to new line
          % Phase 1 Ginput tracker selection


if skipPhase1TMat == 'N'

    counter06_B = 1; %counter indicating the row in the matrix where trackers will be saved (starts at 1, unlike counter 06 which starts at some
frame number)

    for counter06 = phase1Start : phase1End   %nested outer for loop, outer loop goes frame by frame, inner loop sets a ginuput for each tracker up
to the set number of trackers

        %load frame image based on counter06 current value
        currentFrameImageP1 = read(videoReader, counter06); %open the current frame according to counter06, and save the image to a new matrix
specifically designed to hold phase 1 data. the same frame should be called with each loop where the tracker is changed (tracker06_A). Note:
williams code relies on a while loop. This code stays consistant by using a counter instead, either way should work.
        imshow(currentFrameImageP1); %display the image
        axis on; %turn on axis to show the pixels to be selected
        hold on; %save all subsequent point images of ginput tracker locations to the current frame image until hold off is specified
        plot(AvgXP1, AvgYP1,'*', 'Color', 'red') %plot the phase center as a reference
        for counter06_A = 1 : prompt_06_AMat  %inner loop performs a calculation for each tracker in phase 1 up to tracker number specified by
user in prompt_06_A

            %ginput on the image for each tracker and save x,y coordinates to the matrix of the current tracker as specified by the inner loop
            if counter06_A == 1
                plot(tracker1Loc(1:counter06_B, 1), tracker1Loc(1:counter06_B, 2),'*', 'Color', 'Blue') %display all previous point selections for the
current tracker as a reference
                %run ginput and save x,y coordinates to tracker1 matrix in the row specified by counter06_B which should count up from 1 to the
total frame number in that phase for each run of the outer loop
                zoom on; %turn on the zoom feature
                pause %allow for zooming in until button is pressed
                [tracker1Loc(counter06_B, 1), tracker1Loc(counter06_B, 2)] = ginput(1); %allow the user to click on a point. Save that point to
trackerLoc matrix specific to the tracker number in columns 1,2 for x,y  if in phase 1, or in columns 3,4 for x,y if in phase 2.
                tracker1Loc(counter06_B, 1) = round (tracker1Loc(counter06_B, 1)); %update the x coordinate of the tracker to a rounded value
                tracker1Loc(counter06_B, 2) = round (tracker1Loc(counter06_B, 2)); %update the y coordinate of the tracker to a rounded value
                zoom out; %turn off the zoom feature, return to original image
                %point1A = insertMarker(currentFrameImageP1, [tracker1Loc(counter06_B, 1), tracker1Loc(counter06_B, 2)],'+', 'Color', 'green');
%plot the now rounded coordinates
                plot(tracker1Loc(counter06_B, 1), tracker1Loc(counter06_B, 2),'*', 'Color', 'green')
                hold on; %prepare to hold the next set of trackers
            elseif counter06_A == 2 %Tracker 2 --> counter06_B = 2
                plot(tracker2Loc(1:counter06_B, 1), tracker2Loc(1:counter06_B, 2),'*', 'Color', 'Blue')
                %run ginput and save x,y coordinates to tracker2 matrix in the row specified by counter06_B which should count up from 1 to the
total frame number in that phase for each run of the outer loop
                zoom on; %turn on the zoom feature
                pause %allow for zooming in until button is pressed
                [tracker2Loc(counter06_B, 1), tracker2Loc(counter06_B, 2)] = ginput(1); %allow the user to click on a point. Save that point to
trackerLoc matrix specific to the tracker number in columns 1,2 for x,y  if in phase 1, or in columns 3,4 for x,y if in phase 2.
                tracker2Loc(counter06_B, 1) = round (tracker2Loc(counter06_B, 1)); %update the x coordinate of the tracker to a rounded value
                tracker2Loc(counter06_B, 2) = round (tracker2Loc(counter06_B, 2)); %update the y coordinate of the tracker to a rounded value
                zoom out; %turn off the zoom feature, return to original image
                %point1B = insertMarker(currentFrameImageP1, [tracker2Loc(counter06_B, 1), tracker2Loc(counter06_B, 2)],'+', 'Color', 'green');
%plot the now rounded coordinates
                plot(tracker2Loc(counter06_B, 1), tracker2Loc(counter06_B, 2),'*', 'Color', 'green')
                hold on;
            elseif counter06_A == 3 %Tracker 3 --> counter06_B = 3
                plot(tracker3Loc(1:counter06_B, 1), tracker3Loc(1:counter06_B, 2),'*', 'Color', 'Blue')
                %run ginput and save x,y coordinates to tracker3 matrix in the row specified by counter06_B which should count up from 1 to the
total frame number in that phase for each run of the outer loop
                zoom on; %turn on the zoom feature
                pause %allow for zooming in until button is pressed
                [tracker3Loc(counter06_B, 1), tracker3Loc(counter06_B, 2)] = ginput(1); %allow the user to click on a point. Save that point to
trackerLoc matrix specific to the tracker number in columns 1,2 for x,y  if in phase 1, or in columns 3,4 for x,y if in phase 2.
                tracker3Loc(counter06_B, 1) = round (tracker3Loc(counter06_B, 1)); %update the x coordinate of the tracker to a rounded value
```

```matlab
            tracker3Loc(counter06_B, 2) = round (tracker3Loc(counter06_B, 2)); %update the y coordinate of the tracker to a rounded value
            zoom out; %turn off the zoom feature, return to original image
            %point1C = insertMarker(currentFrameImageP1, [tracker3Loc(counter06_B, 1), tracker3Loc(counter06_B, 2)],'+', 'Color', 'green');
%plot the now rounded coordinates
            plot(tracker3Loc(counter06_B, 1), tracker3Loc(counter06_B, 2),'*', 'Color', 'green')
            hold on;
        elseif counter06_A == 4 %Tracker 4 --> counter06_B = 4
            plot(tracker4Loc(1:counter06_B, 1), tracker4Loc(1:counter06_B, 2),'*', 'Color', 'Blue')
            %run ginput and save x,y coordinates to tracker2 matrix in the row specified by counter06_B which should count up from 1 to the
total frame number in that phase for each run of the outer loop
            zoom on; %turn on the zoom feature
            pause %allow for zooming in until button is pressed
            [tracker4Loc(counter06_B, 1), tracker4Loc(counter06_B, 2)] = ginput(1); %allow the user to click on a point. Save that point to
trackerLoc matrix specific to the tracker number in columns 1,2 for x,y  if in phase 1, or in columns 3,4 for x,y if in phase 2.
            tracker4Loc(counter06_B, 1) = round (tracker4Loc(counter06_B, 1)); %update the x coordinate of the tracker to a rounded value
            tracker4Loc(counter06_B, 2) = round (tracker4Loc(counter06_B, 2)); %update the y coordinate of the tracker to a rounded value
            zoom out; %turn off the zoom feature, return to original image
            %point1D = insertMarker(currentFrameImageP1, [tracker4Loc(counter06_B, 1), tracker4Loc(counter06_B, 2)],'+', 'Color', 'green');
%plot the now rounded coordinates
            plot(tracker4Loc(counter06_B, 1), tracker4Loc(counter06_B, 2),'*', 'Color', 'green')
            hold on;
        elseif counter06_A == 5 %Tracker 5 --> counter06_B = 5
            plot(tracker5Loc(1:counter06_B, 1), tracker5Loc(1:counter06_B, 2),'*', 'Color', 'Blue')
            %run ginput and save x,y coordinates to tracker2 matrix in the row specified by counter06_B which should count up from 1 to the
total frame number in that phase for each run of the outer loop
            zoom on; %turn on the zoom feature
            pause %allow for zooming in until button is pressed
            [tracker5Loc(counter06_B, 1), tracker5Loc(counter06_B, 2)] = ginput(1); %allow the user to click on a point. Save that point to
trackerLoc matrix specific to the tracker number in columns 1,2 for x,y  if in phase 1, or in columns 3,4 for x,y if in phase 2.
            tracker5Loc(counter06_B, 1) = round (tracker5Loc(counter06_B, 1)); %update the x coordinate of the tracker to a rounded value
            tracker5Loc(counter06_B, 2) = round (tracker5Loc(counter06_B, 2)); %update the y coordinate of the tracker to a rounded value
            zoom out; %turn off the zoom feature, return to original image
            %point1E = insertMarker(currentFrameImageP1, [tracker5Loc(counter06_B, 1), tracker5Loc(counter06_B, 2)],'+', 'Color', 'green');
%plot the now rounded coordinates
            plot(tracker5Loc(counter06_B, 1), tracker5Loc(counter06_B, 2),'*', 'Color', 'green')
            hold on;
        elseif counter06_A == 6 %Tracker 6 --> counter06_B = 6
            plot(tracker6Loc(1:counter06_B, 1), tracker6Loc(1:counter06_B, 2),'*', 'Color', 'Blue')
            %run ginput and save x,y coordinates to tracker2 matrix in the row specified by counter06_B which should count up from 1 to the
total frame number in that phase for each run of the outer loop
            zoom on; %turn on the zoom feature
            pause %allow for zooming in until button is pressed
            [tracker6Loc(counter06_B, 1), tracker6Loc(counter06_B, 2)] = ginput(1); %allow the user to click on a point. Save that point to
trackerLoc matrix specific to the tracker number in columns 1,2 for x,y  if in phase 1, or in columns 3,4 for x,y if in phase 2.
            tracker6Loc(counter06_B, 1) = round (tracker6Loc(counter06_B, 1)); %update the x coordinate of the tracker to a rounded value
            tracker6Loc(counter06_B, 2) = round (tracker6Loc(counter06_B, 2)); %update the y coordinate of the tracker to a rounded value
            zoom out; %turn off the zoom feature, return to original image
            %point1F = insertMarker(currentFrameImageP1, [tracker6Loc(counter06_B, 1), tracker6Loc(counter06_B, 2)],'+', 'Color', 'green');
%plot the now rounded coordinates
            plot(tracker6Loc(counter06_B, 1), tracker6Loc(counter06_B, 2),'*', 'Color', 'green')
            hold on;
        elseif counter06_A == 7 %Tracker 7 --> counter06_B = 7
            plot(tracker7Loc(1:counter06_B, 1), tracker7Loc(1:counter06_B, 2),'*', 'Color', 'Blue')
            %run ginput and save x,y coordinates to tracker2 matrix in the row specified by counter06_B which should count up from 1 to the
total frame number in that phase for each run of the outer loop
            zoom on; %turn on the zoom feature
            pause %allow for zooming in until button is pressed
            [tracker7Loc(counter06_B, 1), tracker7Loc(counter06_B, 2)] = ginput(1); %allow the user to click on a point. Save that point to
trackerLoc matrix specific to the tracker number in columns 1,2 for x,y  if in phase 1, or in columns 3,4 for x,y if in phase 2.
            tracker7Loc(counter06_B, 1) = round (tracker7Loc(counter06_B, 1)); %update the x coordinate of the tracker to a rounded value
            tracker7Loc(counter06_B, 2) = round (tracker7Loc(counter06_B, 2)); %update the y coordinate of the tracker to a rounded value
            zoom out; %turn off the zoom feature, return to original image
            %point1G = insertMarker(currentFrameImageP1, [tracker7Loc(counter06_B, 1), tracker7Loc(counter06_B, 2)],'+', 'Color', 'green');
%plot the now rounded coordinates
            plot(tracker7Loc(counter06_B, 1), tracker7Loc(counter06_B, 2),'*', 'Color', 'green')
            hold on;
        elseif counter06_A == 8 %Tracker 8 --> counter06_B = 8
            plot(tracker8Loc(1:counter06_B, 1), tracker8Loc(1:counter06_B, 2),'*', 'Color', 'Blue')
            %run ginput and save x,y coordinates to tracker2 matrix in the row specified by counter06_B which should count up from 1 to the
total frame number in that phase for each run of the outer loop
```

```
                zoom on; %turn on the zoom feature
                pause %allow for zooming in until button is pressed
                [tracker8Loc(counter06_B, 1), tracker8Loc(counter06_B, 2)] = ginput(1); %allow the user to click on a point. Save that point to
trackerLoc matrix specific to the tracker number in columns 1,2 for x,y  if in phase 1, or in columns 3,4 for x,y if in phase 2.
                tracker8Loc(counter06_B, 1) = round (tracker8Loc(counter06_B, 1)); %update the x coordinate of the tracker to a rounded value
                tracker8Loc(counter06_B, 2) = round (tracker8Loc(counter06_B, 2)); %update the y coordinate of the tracker to a rounded value
                zoom out; %turn off the zoom feature, return to original image
                %point1H = insertMarker(currentFrameImageP1, [tracker8Loc(counter06_B, 1), tracker8Loc(counter06_B, 2)],'+', 'Color', 'green');
%plot the now rounded coordinates
                plot(tracker8Loc(counter06_B, 1), tracker8Loc(counter06_B, 2),'*', 'Color', 'green')
                hold on;
            elseif counter06_A == 9 %Tracker 9 --> counter06_B = 9
                plot(tracker9Loc(1:counter06_B, 1), tracker9Loc(1:counter06_B, 2),'*', 'Color', 'Blue')
                %run ginput and save x,y coordinates to tracker2 matrix in the row specified by counter06_B which should count up from 1 to the
total frame number in that phase for each run of the outer loop
                zoom on; %turn on the zoom feature
                pause %allow for zooming in until button is pressed
                [tracker9Loc(counter06_B, 1), tracker9Loc(counter06_B, 2)] = ginput(1); %allow the user to click on a point. Save that point to
trackerLoc matrix specific to the tracker number in columns 1,2 for x,y  if in phase 1, or in columns 3,4 for x,y if in phase 2.
                tracker9Loc(counter06_B, 1) = round (tracker9Loc(counter06_B, 1)); %update the x coordinate of the tracker to a rounded value
                tracker9Loc(counter06_B, 2) = round (tracker9Loc(counter06_B, 2)); %update the y coordinate of the tracker to a rounded value
                zoom out; %turn off the zoom feature, return to original image
                %point1I = insertMarker(currentFrameImageP1, [tracker9Loc(counter06_B, 1), tracker9Loc(counter06_B, 2)],'+', 'Color', 'green');
%plot the now rounded coordinates
                plot(tracker9Loc(counter06_B, 1), tracker9Loc(counter06_B, 2),'*', 'Color', 'green')
                hold on;
            elseif counter06_A == 10 %Tracker 10 --> counter06_B = 10
                plot(tracker10Loc(1:counter06_B, 1), tracker10Loc(1:counter06_B, 2),'*', 'Color', 'Blue')
                %run ginput and save x,y coordinates to tracker2 matrix in the row specified by counter06_B which should count up from 1 to the
total frame number in that phase for each run of the outer loop
                zoom on; %turn on the zoom feature
                pause %allow for zooming in until button is pressed
                [tracker10Loc(counter06_B, 1), tracker10Loc(counter06_B, 2)] = ginput(1); %allow the user to click on a point. Save that point to
trackerLoc matrix specific to the tracker number in columns 1,2 for x,y  if in phase 1, or in columns 3,4 for x,y if in phase 2.
                tracker10Loc(counter06_B, 1) = round (tracker10Loc(counter06_B, 1)); %update the x coordinate of the tracker to a rounded value
                tracker10Loc(counter06_B, 2) = round (tracker10Loc(counter06_B, 2)); %update the y coordinate of the tracker to a rounded value
                zoom out; %turn off the zoom feature, return to original image
                %point1J = insertMarker(currentFrameImageP1, [tracker10Loc(counter06_B, 1), tracker10Loc(counter06_B, 2)],'+', 'Color', 'green');
%plot the now rounded coordinates
                plot(tracker10Loc(counter06_B, 1), tracker10Loc(counter06_B, 2),'*', 'Color', 'green')
                hold on;
            end %end ginput point selection for the tracker number specified by counter06_A.
            hold on; %verify that the point image for the previous tracker is saved when moving on to the next tracker in the current frame
        end %end the for loop which repeats the ginput loop (end above this one) for each tracker until all trackers in phase 1 have been defined.

    %Save the current frame image now updated to show all trackers  (pointA, pointB .... pointI) for phase 1.
        % create a file name variably with the sprintf command
        baseFileName = sprintf('Phase1_frame_%d.jpg', counter06_B);
        % Specify some particular, specific folder:
        fullFileName = fullfile('C:\Users\Name\Desktop\Matlab\Photos', baseFileName);
        % Export the open image
        exportgraphics(gcf, fullFileName); % Using export_fig instead of saveas.
        %Close the image
        close all; %close the imageviewer frame so the next loop can start over at image 1
    hold off; %once all trackers have been saved to the current frame, allow the next frame to clear the tracker images from the previous frame
and start fresh
        counter06_B = counter06_B + 1; %counter06_B saves the current frame number starting with 1 and increments upward through all frames:
at this point the inner loop has already ended (all tracker matrices have been updated with one row (row1) of ginput coordinates), the counter
indicating current frame number in the phase is incremented so that the next set of coordinates will be saved in row 2 of each matrix
        %by incrementing the counter at this point, the for loop can be rerun covering all trackers on the next video image frame, and save the
results to the next row in each tracker matrix

    end %ends the phase 1 for loop which ensured that all frames were inspected, with each frame given its own row of data in the tracker matrices
(up to 10 tracker matrices depending on the number of trackers) (each matrix haveing the same number of rows as number of frames, +1 to
include data for the 0th frame)

end


    fprintf('Phase 1 Tracker Selection Complete! \n') %notify user of successful ginput, \n skips to new line
```

```matlab
% Phase 2 Ginput tracker selection
fprintf('Begin Phase 2 Tracker Selection! \n')  %notify user of start ginput, \n skips to new line



counter07_B = 1; %counter indicating the row in the matrix where trackers will be saved (starts at 1, unlike counter 06 which starts at some
frame number)

for counter07 = phase2Start : phase2End   %nested outer for loop, outer loop goes frame by frame, inner loop sets a ginuput for each tracker up
to the set number of trackers

    %load frame image based on counter06 current value
    currentFrameImageP2 = read(videoReader, counter07); %open the current frame according to counter06, and save the image to a new matrix
specifically designed to hold phase 1 data. the same frame should be called with each loop where the tracker is changed (tracker06_A). Note:
williams code relies on a while loop. This code stays consistant by using a counter instead, either way should work.
    imshow(currentFrameImageP2); %display the image
    axis on; %turn on axis to show the pixels to be selected
    hold on;
    plot(AvgXP2, AvgYP2,'*', 'Color', 'red') %plot the phase center as a reference
    for counter07_A = 1 : prompt_06_BMat  %inner loop performs a calculation for each tracker in phase 1 up to tracker number specified by
user in prompt_06_B

        %ginput on the image for each tracker and save x,y coordinates to the matrix of the current tracker as specified by the inner loop
        if counter07_A == 1
            plot(tracker1Loc(1:counter07_B, 3), tracker1Loc(1:counter07_B, 4),'*', 'Color', 'Blue') %plot all previous points for tracker 1 as a
reference
            %run ginput and save x,y coordinates to tracker1 matrix in the row specified by counter06_B which should count up from 1 to the
total frame number in that phase for each run of the outer loop
            zoom on; %turn on the zoom feature
            pause %allow for zooming in until button is pressed
            [tracker1Loc(counter07_B, 3), tracker1Loc(counter07_B, 4)] = ginput(1); %allow the user to click on a point. Save that point to
trackerLoc matrix specific to the tracker number in columns 1,2 for x,y  if in phase 1, or in columns 3,4 for x,y if in phase 2.
            tracker1Loc(counter07_B, 3) = round (tracker1Loc(counter07_B, 3)); %update the x coordinate of the tracker to a rounded value
            tracker1Loc(counter07_B, 4) = round (tracker1Loc(counter07_B, 4)); %update the y coordinate of the tracker to a rounded value
            zoom out; %turn off the zoom feature, return to original image
            %point2A = insertMarker(currentFrameImageP2, [tracker1Loc(counter07_B, 3), tracker1Loc(counter07_B, 4)],'+', 'Color', 'green');
%plot the now rounded coordinates
            plot(tracker1Loc(counter07_B, 3), tracker1Loc(counter07_B, 4),'*', 'Color', 'green')
            hold on;
        elseif counter07_A == 2 %Tracker 2 --> counter07_B = 2
            plot(tracker2Loc(1:counter07_B, 3), tracker2Loc(1:counter07_B, 4),'*', 'Color', 'Blue')
            %run ginput and save x,y coordinates to tracker2 matrix in the row specified by counter06_B which should count up from 1 to the
total frame number in that phase for each run of the outer loop
            zoom on; %turn on the zoom feature
            pause %allow for zooming in until button is pressed
            [tracker2Loc(counter07_B, 3), tracker2Loc(counter07_B, 4)] = ginput(1); %allow the user to click on a point. Save that point to
trackerLoc matrix specific to the tracker number in columns 1,2 for x,y  if in phase 1, or in columns 3,4 for x,y if in phase 2.
            tracker2Loc(counter07_B, 3) = round (tracker2Loc(counter07_B, 3)); %update the x coordinate of the tracker to a rounded value
            tracker2Loc(counter07_B, 4) = round (tracker2Loc(counter07_B, 4)); %update the y coordinate of the tracker to a rounded value
            zoom out; %turn off the zoom feature, return to original image
            %point2B = insertMarker(currentFrameImageP2, [tracker2Loc(counter07_B, 3), tracker2Loc(counter07_B, 4)],'+', 'Color', 'green');
%plot the now rounded coordinates
            plot(tracker2Loc(counter07_B, 3), tracker2Loc(counter07_B, 4),'*', 'Color', 'green')
            hold on;
        elseif counter07_A == 3 %Tracker 3 --> counter07_B = 3
            plot(tracker3Loc(1:counter07_B, 3), tracker3Loc(1:counter07_B, 4),'*', 'Color', 'Blue')
            %run ginput and save x,y coordinates to tracker3 matrix in the row specified by counter06_B which should count up from 1 to the
total frame number in that phase for each run of the outer loop
            zoom on; %turn on the zoom feature
            pause %allow for zooming in until button is pressed
            [tracker3Loc(counter07_B, 3), tracker3Loc(counter07_B, 4)] = ginput(1); %allow the user to click on a point. Save that point to
trackerLoc matrix specific to the tracker number in columns 1,2 for x,y  if in phase 1, or in columns 3,4 for x,y if in phase 2.
            tracker3Loc(counter07_B, 3) = round (tracker3Loc(counter07_B, 3)); %update the x coordinate of the tracker to a rounded value
            tracker3Loc(counter07_B, 4) = round (tracker3Loc(counter07_B, 4)); %update the y coordinate of the tracker to a rounded value
            zoom out; %turn off the zoom feature, return to original image
            %point2C = insertMarker(currentFrameImageP2, [tracker3Loc(counter07_B, 3), tracker3Loc(counter07_B, 4)],'+', 'Color', 'green');
%plot the now rounded coordinates
            plot(tracker3Loc(counter07_B, 3), tracker3Loc(counter07_B, 4),'*', 'Color', 'green')
            hold on;
```

```matlab
            elseif counter07_A == 4 %Tracker 4 --> counter07_B = 4
                plot(tracker4Loc(1:counter07_B, 3), tracker4Loc(1:counter07_B, 4),'*', 'Color', 'Blue')
                %run ginput and save x,y coordinates to tracker2 matrix in the row specified by counter06_B which should count up from 1 to the
total frame number in that phase for each run of the outer loop
                zoom on; %turn on the zoom feature
                pause %allow for zooming in until button is pressed
                [tracker4Loc(counter07_B, 3), tracker4Loc(counter07_B, 4)] = ginput(1); %allow the user to click on a point. Save that point to
trackerLoc matrix specific to the tracker number in columns 1,2 for x,y  if in phase 1, or in columns 3,4 for x,y  if in phase 2.
                tracker4Loc(counter07_B, 3) = round (tracker4Loc(counter07_B, 3)); %update the x coordinate of the tracker to a rounded value
                tracker4Loc(counter07_B, 4) = round (tracker4Loc(counter07_B, 4)); %update the y coordinate of the tracker to a rounded value
                zoom out; %turn off the zoom feature, return to original image
                %point2D = insertMarker(currentFrameImageP2, [tracker4Loc(counter07_B, 3), tracker4Loc(counter07_B, 4)],'+', 'Color', 'green');
%plot the now rounded coordinates
                plot(tracker4Loc(counter07_B, 3), tracker4Loc(counter07_B, 4),'*', 'Color', 'green')
                hold on;
            elseif counter07_A == 5 %Tracker 5 --> counter07_B = 5
                plot(tracker5Loc(1:counter07_B, 3), tracker5Loc(1:counter07_B, 4),'*', 'Color', 'Blue')
                %run ginput and save x,y coordinates to tracker2 matrix in the row specified by counter06_B which should count up from 1 to the
total frame number in that phase for each run of the outer loop
                zoom on; %turn on the zoom feature
                pause %allow for zooming in until button is pressed
                [tracker5Loc(counter07_B, 3), tracker5Loc(counter07_B, 4)] = ginput(1); %allow the user to click on a point. Save that point to
trackerLoc matrix specific to the tracker number in columns 1,2 for x,y  if in phase 1, or in columns 3,4 for x,y  if in phase 2.
                tracker5Loc(counter07_B, 3) = round (tracker5Loc(counter07_B, 3)); %update the x coordinate of the tracker to a rounded value
                tracker5Loc(counter07_B, 4) = round (tracker5Loc(counter07_B, 4)); %update the y coordinate of the tracker to a rounded value
                zoom out; %turn off the zoom feature, return to original image
                %point2E = insertMarker(currentFrameImageP2, [tracker5Loc(counter07_B, 3), tracker5Loc(counter07_B, 4)],'+', 'Color', 'green');
%plot the now rounded coordinates
                plot(tracker5Loc(counter07_B, 3), tracker5Loc(counter07_B, 4),'*', 'Color', 'green')
                hold on;
            elseif counter07_A == 6 %Tracker 6 --> counter07_B = 6
                plot(tracker6Loc(1:counter07_B, 3), tracker6Loc(1:counter07_B, 4),'*', 'Color', 'Blue')
                %run ginput and save x,y coordinates to tracker2 matrix in the row specified by counter06_B which should count up from 1 to the
total frame number in that phase for each run of the outer loop
                zoom on; %turn on the zoom feature
                pause %allow for zooming in until button is pressed
                [tracker6Loc(counter07_B, 3), tracker6Loc(counter07_B, 4)] = ginput(1); %allow the user to click on a point. Save that point to
trackerLoc matrix specific to the tracker number in columns 1,2 for x,y  if in phase 1, or in columns 3,4 for x,y  if in phase 2.
                tracker6Loc(counter07_B, 3) = round (tracker6Loc(counter07_B, 3)); %update the x coordinate of the tracker to a rounded value
                tracker6Loc(counter07_B, 4) = round (tracker6Loc(counter07_B, 4)); %update the y coordinate of the tracker to a rounded value
                zoom out; %turn off the zoom feature, return to original image
                %point2F = insertMarker(currentFrameImageP2, [tracker6Loc(counter07_B, 3), tracker6Loc(counter07_B, 4)],'+', 'Color', 'green');
%plot the now rounded coordinates
                plot(tracker6Loc(counter07_B, 3), tracker6Loc(counter07_B, 4),'*', 'Color', 'green')
                hold on;
            elseif counter07_A == 7 %Tracker 7 --> counter07_B = 7
                plot(tracker7Loc(1:counter07_B, 3), tracker7Loc(1:counter07_B, 4),'*', 'Color', 'Blue')
                %run ginput and save x,y coordinates to tracker2 matrix in the row specified by counter06_B which should count up from 1 to the
total frame number in that phase for each run of the outer loop
                zoom on; %turn on the zoom feature
                pause %allow for zooming in until button is pressed
                [tracker7Loc(counter07_B, 3), tracker7Loc(counter07_B, 4)] = ginput(1); %allow the user to click on a point. Save that point to
trackerLoc matrix specific to the tracker number in columns 1,2 for x,y  if in phase 1, or in columns 3,4 for x,y  if in phase 2.
                tracker7Loc(counter07_B, 3) = round (tracker7Loc(counter07_B, 3)); %update the x coordinate of the tracker to a rounded value
                tracker7Loc(counter07_B, 4) = round (tracker7Loc(counter07_B, 4)); %update the y coordinate of the tracker to a rounded value
                zoom out; %turn off the zoom feature, return to original image
                %point2G = insertMarker(currentFrameImageP2, [tracker7Loc(counter07_B, 3), tracker7Loc(counter07_B, 4)],'+', 'Color', 'green');
%plot the now rounded coordinates
                plot(tracker7Loc(counter07_B, 3), tracker7Loc(counter07_B, 4),'*', 'Color', 'green')
                hold on;
            elseif counter07_A == 8 %Tracker 8 --> counter07_B = 8
                plot(tracker8Loc(1:counter07_B, 3), tracker8Loc(1:counter07_B, 4),'*', 'Color', 'Blue')
                %run ginput and save x,y coordinates to tracker2 matrix in the row specified by counter06_B which should count up from 1 to the
total frame number in that phase for each run of the outer loop
                zoom on; %turn on the zoom feature
                pause %allow for zooming in until button is pressed
                [tracker8Loc(counter07_B, 3), tracker8Loc(counter07_B, 4)] = ginput(1); %allow the user to click on a point. Save that point to
trackerLoc matrix specific to the tracker number in columns 1,2 for x,y  if in phase 1, or in columns 3,4 for x,y  if in phase 2.
                tracker8Loc(counter07_B, 3) = round (tracker8Loc(counter07_B, 3)); %update the x coordinate of the tracker to a rounded value
                tracker8Loc(counter07_B, 4) = round (tracker8Loc(counter07_B, 4)); %update the y coordinate of the tracker to a rounded value
```

```matlab
            zoom out; %turn off the zoom feature, return to original image
            %point2H = insertMarker(currentFrameImageP2, [tracker8Loc(counter07_B, 3), tracker8Loc(counter07_B, 4)],'+', 'Color', 'green');
%plot the now rounded coordinates
            plot(tracker8Loc(counter07_B, 3), tracker8Loc(counter07_B, 4),'*', 'Color', 'green')
            hold on;
        elseif counter07_A == 9 %Tracker 9 --> counter07_B = 9
            plot(tracker9Loc(1:counter07_B, 3), tracker9Loc(1:counter07_B, 4),'*', 'Color', 'Blue')
            %run ginput and save x,y coordinates to tracker2 matrix in the row specified by counter06_B which should count up from 1 to the
total frame number in that phase for each run of the outer loop
            zoom on; %turn on the zoom feature
            pause %allow for zooming in until button is pressed
            [tracker9Loc(counter07_B, 3), tracker9Loc(counter07_B, 4)] = ginput(1); %allow the user to click on a point. Save that point to
trackerLoc matrix specific to the tracker number in columns 1,2 for x,y  if in phase 1, or in columns 3,4 for x,y if in phase 2.
            tracker9Loc(counter07_B, 3) = round (tracker9Loc(counter07_B, 3)); %update the x coordinate of the tracker to a rounded value
            tracker9Loc(counter07_B, 4) = round (tracker9Loc(counter07_B, 4)); %update the y coordinate of the tracker to a rounded value
            zoom out; %turn off the zoom feature, return to original image
            %point2I = insertMarker(currentFrameImageP2, [tracker9Loc(counter07_B, 3), tracker9Loc(counter07_B, 4)],'+', 'Color', 'green');
%plot the now rounded coordinates
            plot(tracker9Loc(counter07_B, 3), tracker9Loc(counter07_B, 4),'*', 'Color', 'green')
            hold on;
        elseif counter07_A == 10 %Tracker 10 --> counter07_B = 10
            plot(tracker10Loc(1:counter07_B, 3), tracker10Loc(1:counter07_B, 4),'*', 'Color', 'Blue')
            %run ginput and save x,y coordinates to tracker2 matrix in the row specified by counter06_B which should count up from 1 to the
total frame number in that phase for each run of the outer loop
            zoom on; %turn on the zoom feature
            pause %allow for zooming in until button is pressed
            [tracker10Loc(counter07_B, 3), tracker10Loc(counter07_B, 4)] = ginput(1); %allow the user to click on a point. Save that point to
trackerLoc matrix specific to the tracker number in columns 1,2 for x,y  if in phase 1, or in columns 3,4 for x,y if in phase 2.
            tracker10Loc(counter07_B, 3) = round (tracker10Loc(counter07_B, 3)); %update the x coordinate of the tracker to a rounded value
            tracker10Loc(counter07_B, 4) = round (tracker10Loc(counter07_B, 4)); %update the y coordinate of the tracker to a rounded value
            zoom out; %turn off the zoom feature, return to original image
            %point2J = insertMarker(currentFrameImageP2, [tracker10Loc(counter07_B, 3), tracker10Loc(counter07_B, 4)],'+', 'Color', 'green');
%plot the now rounded coordinates
            plot(tracker10Loc(counter07_B, 3), tracker10Loc(counter07_B, 4),'*', 'Color', 'green')
            hold on;
        end %end ginput point selection for the tracker number specified by counter06_A.
        hold on; %verify the current tracker image has been saved before moving on to the next tracker (same frame only, remove the hold to
clear all trackers before the start of the next frame)
    end %end the for loop which repeats the ginput loop (end above this one) for each tracker until all trackers in phase 1 have been defined.

    %Save the current frame image now updated to show all trackers  (pointA, pointB .... pointI) for phase 1.
    % create a file name variably with the sprintf command
    baseFileName2 = sprintf('Phase2_frame_%d.jpg', counter07_B);
    % Specify some particular, specific folder:
    fullFileName2 = fullfile('C:\Users\Name\Desktop\Matlab\Photos', baseFileName2);
    % Export the open image
    exportgraphics(gcf, fullFileName2); % Using export_fig instead of saveas.
    %Close the image
    close all; %close the imageviewer frame so the next loop can start over at image 1
    hold off;
    counter07_B = counter07_B + 1; %counter06_B saves the current frame number starting with 1 and increments upward through all frames:
at this point the inner loop has already ended (all tracker matrices have been updated with one row (row1) of ginput coordinates), the counter
indicating current frame number in the phase is incremented so that the next set of coordinates will be saved in row 2 of each matrix
    %by incrementing the counter at this point, the for loop can be rerun covering all trackers on the next video image frame, and save the
results to the next row in each tracker matrix

end %ends the phase 1 for loop which ensured that all frames were inspected, with each frame given its own row of data in the tracker matrices
(up to 10 tracker matrices depending on the number of trackers) (each matrix haveing the same number of rows as number of frames, +1 to
include data for the 0th frame)


fprintf('Phase 2 Tracker Selection Complete! \n') %notify user of successful ginput, \n skips to new line
% End Phase 1 and 2 Ginput tracker selection
```

```matlab
    close; %close the imageviewer frame
end

%% Prompt 06 Results


fprintf('Check Desktop --> Matlab --> Photos for pictures of your point selections \n');

    for counter05 = 1 : maxNumTrackersMat  %this loop will create tracker matrices for each tracker with the number of rows specified by the
number of frames
        if counter05 == 1
            fprintf('Tracker 1 point selections for phases 1 and 2 \n');
            disp(tracker1Loc)
        elseif counter05 == 2
            fprintf('Tracker 2 point selections for phases 1 and 2 \n');
            disp(tracker2Loc)
        elseif counter05 == 3
            fprintf('Tracker 3 point selections for phases 1 and 2 \n');
            disp(tracker3Loc)
        elseif counter05 == 4
            fprintf('Tracker 4 point selections for phases 1 and 2 \n');
            disp(tracker4Loc)
        elseif counter05 == 5
            fprintf('Tracker 5 point selections for phases 1 and 2 \n');
            disp(tracker5Loc)
        elseif counter05 == 6
            fprintf('Tracker 6 point selections for phases 1 and 2 \n');
            disp(tracker6Loc)
        elseif counter05 == 7
            fprintf('Tracker 7 point selections for phases 1 and 2 \n');
            disp(tracker7Loc)
        elseif counter05 == 8
            fprintf('Tracker 8 point selections for phases 1 and 2 \n');
            disp(tracker8Loc)
        elseif counter05 == 9
            fprintf('Tracker 9 point selectionsfor phases 1 and 2 \n');
            disp(tracker9Loc)
        elseif counter05 == 10
            fprintf('Tracker 10 point selections for phases 1 and 2 \n');
            disp(tracker10Loc)
        end % end the if statement
    end % end the for loop

    fprintf('Prompt 06 Complete! \n')


%% Calculations

%called in variables
    %phase center--> AvgXP1, AvgYP1, AvgXP2, AvgYP2
    %circle center--> centersAverage(1,2)
    %tracker points--> tracker1Loc .... tracker10Loc
    %maxNumTrackersMat used in for for loop
    %number of frames in phase 1 (frameTotPhase1)
    %number of frames in phase 2 (frameTotPhase2)

%initialized variables
    %counter08 = 0; %Dont need to initialize variables found in a for loop
    %because the phase centers are only one frame their information about displacement from center can be saved in a single matrix to save space
        delXYZCenter = zeros(2,3); %matrix of displacements between the phase center and the circle center. one row for each of the two phases,
and 3 columns for x, y, z displacements.
    %create a matrix for final matrix coordinates using the z coordinate from delXYZCenter and AvgPX, AvgPY coordinates
        phaseInitialCoord = zeros(2,3);
    %displacement of x and y coordinates ofeach tracker relative to the circle center are necessary to computing the 3-D z coordinate of those
points but arent useful for much else
        delXP1 = zeros(maxFrames, 10); % x=frame number, y=tracker number. In this matrix the displacement between the circle center
coordinate and the tracker point is stored
        delYP1 = zeros(maxFrames, 10);
        delXP2 = zeros(maxFrames, 10);
        delYP2 = zeros(maxFrames, 10);
```

%use displacements saved to the matrices above to fill in the following z coordinate matrix
    delZP1 = zeros(maxFrames, 10); %matrix of z displacements (assuming circle center is at z=0) to be computed for phase 1 trackers. rows
being arbitrarily large enough to hold the z coordinated for up to max number of frames, and 10 columns representing the maximum of 10
trackers that the user might select.
    delZP2 = zeros(maxFrames, 10); %matrix of z displacements (assuming circle center is at z=0) to be computed for phase 2 trackers
    %displacements of x,y,z will also need to be stored between every two frames of the same tracker (phase 1 and phase 2) with the following
columns:(column1 = xP1, column2 = yP1, column3 = zP1, column4 = xP2, column5 = yP2, column6 = zP2)
    delT01 = zeros(maxFrames, 6); %x = frame number, y = coordinate type (x,y,z)p1 (x,y,z)p2
    delT02 = zeros(maxFrames, 6);
    delT03 = zeros(maxFrames, 6);
    delT04 = zeros(maxFrames, 6);
    delT05 = zeros(maxFrames, 6);
    delT06 = zeros(maxFrames, 6);
    delT07 = zeros(maxFrames, 6);
    delT08 = zeros(maxFrames, 6);
    delT09 = zeros(maxFrames, 6);
    delT10 = zeros(maxFrames, 6);
    %Finally Euclidean Distance will be stored to a matrix with rows for each frame and columns for each tracker
    euclideanDistP1 = zeros(maxFrames, 10);   %records euclidean distances between frames for phase 1
    euclideanDistP2 = zeros(maxFrames, 10);   %records euclidean distances between frames for phase 2
    %Then the Euclidean Distances for each tracker will be summed up for all previous frames to current providing total distance from phase
center at the current frame.
    euclideanSumP1 = zeros(frameTotPhase1,prompt_06_AMat);
    euclideanSumP2 = zeros(frameTotPhase2,prompt_06_BMat);

%Compute the Z coordinate of both phase centers
    %phase 1 Z coordinate
    delXYZCenter(1,1) = AvgXP1(1,1) - centersAverage(1,1); %subtract the phase center
    delXYZCenter(1,2) = centersAverage(1,2) - AvgYP1(1,1); %y coordinate is subtracted from the center y coordinate because of the image
scale being inverted with the 0 starting at the top for the vertical axis
    delXYZCenter(1,3) = sqrt((centersAverage(1,3).^2)-(delXYZCenter(1,1).^2)-(delXYZCenter(1,2).^2)); % trackers phase 1 z coordinate
saved to the first row, third column = square root of (radius squared - change in x squared - change in y squared)
    %phase 2 Z coordinate
    delXYZCenter(2,1) = AvgXP2(1,1) - centersAverage(1,1);
    delXYZCenter(2,2) = centersAverage(1,2) - AvgYP2(1,1); %y coordinate is subtracted from the center y coordinate because of the image
scale being inverted with the 0 starting at the top for the vertical axis
    delXYZCenter(2,3) = sqrt((centersAverage(1,3).^2)-(delXYZCenter(2,1).^2)-(delXYZCenter(2,2).^2)); % trackers phase 2 z coordinate
saved to the second row third column = square root of (radius squared - change in x squared - change in y squared)
    %Finally Compile Phase coordinates (x,y,z) now that z has been obtained
        %Phase1 Row 1 columns (x,y,z)
        phaseInitialCoord(1,1) = AvgXP1(1,1);
        phaseInitialCoord(1,2) = AvgYP1(1,1);
        phaseInitialCoord(1,3) = delXYZCenter(1,3);
        %Phase2 Row 2 columns (x,y,z)
        phaseInitialCoord(2,1) = AvgXP2(1,1);
        phaseInitialCoord(2,2) = AvgYP2(1,1);
        phaseInitialCoord(2,3) = delXYZCenter(2,3);

%now do the same for each tracker and compare the x,y,z with x,y,z of their respective phase center to obtain Euclidean Distance. (repeat for
phases 1 and 2
for counter08 = 1: maxNumTrackersMat %increment counter08 starting from 1 until it reaches the number of trackers used. run one calculation
for each tracker,and save to a different row each time.

    if counter08 == 1 %for tracker #1 run this line of code if at least one tracker was used, perform a calculation on data from the first tracker

        %find translation distance between average center and phase 1 for each frame (row) within tracker 1 matrix (x and y data saved in columns 1
and 2)
        %from williams translation of POI code, this step allows delta x, delta y, z between the point and the circle center to be determined

        %for 1 to total number of frames in phase 1, calculate the zed coordinate for tracker 1 for each frame and record as a row vector
        for counter08_A = 1: frameTotPhase1
            delXP1(counter08_A, 1) = tracker1Loc(counter08_A,1) - centersAverage(1,1);
            delYP1(counter08_A, 1) = centersAverage(1,2) - tracker1Loc(counter08_A,2);
            delZP1(counter08_A, 1) = sqrt((centersAverage(1,3).^2)-(delXP1(counter08_A, 1).^2)-(delYP1(counter08_A, 1).^2));
        end
            delT01(1,1) =   tracker1Loc(1,1)   - phaseInitialCoord(1,1);
            delT01(1,2) =   tracker1Loc(1,2)   - phaseInitialCoord(1,2);
            delT01(1,3) =   delZP1(1,1)   - phaseInitialCoord(1,3);

```matlab
        euclideanDistP1(1, 1) = sqrt( (delT01(1, 1).^2) + (delT01(1, 2).^2) + (delT01(1, 3).^2)); %save the first row of eulcidean distance as the
net distance traveled by tracker 1 from the phase center in frame 1 to the tracker point in frame 1.
        for counter08_B = 2: frameTotPhase1
            delT01(counter08_B, 1) = tracker1Loc(counter08_B, 1) - tracker1Loc(counter08_B-1, 1);
            delT01(counter08_B, 2) = tracker1Loc(counter08_B, 2) - tracker1Loc(counter08_B-1, 2);
            delT01(counter08_B, 3) =  delZP1(counter08_B, 1)  -  delZP1(counter08_B-1, 1);
            euclideanDistP1(counter08_B, 1) = sqrt( (delT01(counter08_B, 1).^2) + (delT01(counter08_B, 2).^2) + (delT01(counter08_B, 3).^2));
%save the second row onward (frames) of Euclidean Distances as the distance between the tracker and the same tracker in the previous frame
        end


        for counter08_C = 1: frameTotPhase2
            delXP2(counter08_C, 1) = tracker1Loc(counter08_C,3) - centersAverage(1,1);
            delYP2(counter08_C, 1) = centersAverage(1,2) - tracker1Loc(counter08_C,4);
            delZP2(counter08_C, 1) = sqrt((centersAverage(1,3).^2)-(delXP2(counter08_C, 1).^2)-(delYP2(counter08_C, 1).^2));
        end
            delT01(1,4) =   tracker1Loc(1,3)   - phaseInitialCoord(2,1);
            delT01(1,5) =   tracker1Loc(1,4)   - phaseInitialCoord(2,2);
            delT01(1,6) =      delZP2(1,1)      - phaseInitialCoord(2,3);
            euclideanDistP2(1, 1) = sqrt( (delT01(1, 4).^2) + (delT01(1, 5).^2) + (delT01(1, 6).^2));
        for counter08_D = 2: frameTotPhase2
            delT01(counter08_D, 4) = tracker1Loc(counter08_D, 3) - tracker1Loc(counter08_D-1, 3);
            delT01(counter08_D, 5) = tracker1Loc(counter08_D, 4) - tracker1Loc(counter08_D-1, 4);
            delT01(counter08_D, 6) =  delZP2(counter08_D, 1)  -  delZP2(counter08_D-1, 1);
            euclideanDistP2(counter08_D, 1) = sqrt( (delT01(counter08_D, 4).^2) + (delT01(counter08_D, 5).^2) + (delT01(counter08_D, 6).^2));
        end

    elseif counter08 == 2 %in the second loop, counter 5 will be incremented if at least 2 trackers were used. run this calculation for the second
tracker if so
        for counter08_E = 1: frameTotPhase1
            delXP1(counter08_E, 2) = tracker2Loc(counter08_E,1) - centersAverage(1,1);
            delYP1(counter08_E, 2) = centersAverage(1,2) - tracker2Loc(counter08_E,2);
            delZP1(counter08_E, 2) = sqrt((centersAverage(1,3).^2)-(delXP1(counter08_E, 2).^2)-(delYP1(counter08_E, 2).^2));
        end
            delT02(1,1) =   tracker2Loc(1,1)   - phaseInitialCoord(1,1);
            delT02(1,2) =   tracker2Loc(1,2)   - phaseInitialCoord(1,2);
            delT02(1,3) =      delZP1(1,2)      - phaseInitialCoord(1,3);
            euclideanDistP1(1, 2) = sqrt( (delT02(1, 1).^2) + (delT02(1, 2).^2) + (delT02(1, 3).^2));
        for counter08_F = 2: frameTotPhase1
            delT02(counter08_F, 1) = tracker2Loc(counter08_F, 1) - tracker2Loc(counter08_F-1, 1);
            delT02(counter08_F, 2) = tracker2Loc(counter08_F, 2) - tracker2Loc(counter08_F-1, 2);
            delT02(counter08_F, 3) =  delZP1(counter08_F, 2)  -  delZP1(counter08_F-1, 2);
            euclideanDistP1(counter08_F, 2) = sqrt( (delT02(counter08_F, 1).^2) + (delT02(counter08_F, 2).^2) + (delT02(counter08_F, 3).^2));
        end


        for counter08_G = 1: frameTotPhase2
            delXP2(counter08_G, 2) = tracker2Loc(counter08_G,3) - centersAverage(1,1);
            delYP2(counter08_G, 2) = centersAverage(1,2) - tracker2Loc(counter08_G,4);
            delZP2(counter08_G, 2) = sqrt((centersAverage(1,3).^2)-(delXP2(counter08_G, 2).^2)-(delYP2(counter08_G, 2).^2));
        end
            delT02(1,4) =   tracker2Loc(1,3)   - phaseInitialCoord(2,1);
            delT02(1,5) =   tracker2Loc(1,4)   - phaseInitialCoord(2,2);
            delT02(1,6) =      delZP2(1,2)      - phaseInitialCoord(2,3);
            euclideanDistP2(1, 2) = sqrt( (delT02(1, 4).^2) + (delT02(1, 5).^2) + (delT02(1, 6).^2));
        for counter08_H = 2: frameTotPhase2
            delT02(counter08_H, 4) = tracker2Loc(counter08_H, 3) - tracker2Loc(counter08_H-1, 3);
            delT02(counter08_H, 5) = tracker2Loc(counter08_H, 4) - tracker2Loc(counter08_H-1, 4);
            delT02(counter08_H, 6) =  delZP2(counter08_H, 2)  -  delZP2(counter08_H-1, 1);
            euclideanDistP2(counter08_H, 2) = sqrt( (delT02(counter08_H, 4).^2) + (delT02(counter08_H, 5).^2) + (delT02(counter08_H, 6).^2));
        end

    elseif counter08 == 3 % run this line of code on the third loop if a third tracker was used
        for counter08_I = 1: frameTotPhase1
            delXP1(counter08_I, 3) = tracker3Loc(counter08_I,1) - centersAverage(1,1);
            delYP1(counter08_I, 3) = centersAverage(1,2) - tracker3Loc(counter08_I,2);
            delZP1(counter08_I, 3) = sqrt((centersAverage(1,3).^2)-(delXP1(counter08_I, 3).^2)-(delYP1(counter08_I, 3).^2));
        end
```

```matlab
        delT03(1,1) =    tracker3Loc(1,1)    - phaseInitialCoord(1,1);
        delT03(1,2) =    tracker3Loc(1,2)    - phaseInitialCoord(1,2);
        delT03(1,3) =     delZP1(1,3)         - phaseInitialCoord(1,3);
        euclideanDistP1(1, 3) = sqrt( (delT03(1, 1).^2) + (delT03(1, 2).^2) + (delT03(1, 3).^2));
    for counter08_J = 2: frameTotPhase1
        delT03(counter08_J, 1) = tracker3Loc(counter08_J, 1) - tracker3Loc(counter08_J-1, 1);
        delT03(counter08_J, 2) = tracker3Loc(counter08_J, 2) - tracker3Loc(counter08_J-1, 2);
        delT03(counter08_J, 3) =  delZP1(counter08_J, 3)    -   delZP1(counter08_J-1, 3);
        euclideanDistP1(counter08_J, 3) = sqrt( (delT03(counter08_J, 1).^2) + (delT03(counter08_J, 2).^2) + (delT03(counter08_J, 3).^2));
    end


    for counter08_K = 1: frameTotPhase2
        delXP2(counter08_K, 3) = tracker3Loc(counter08_K,3) - centersAverage(1,1);
        delYP2(counter08_K, 3) = centersAverage(1,2) - tracker3Loc(counter08_K,4);
        delZP2(counter08_K, 3) = sqrt((centersAverage(1,3).^2)-(delXP2(counter08_K, 3).^2)-(delYP2(counter08_K, 3).^2));
    end
        delT03(1,4) =    tracker3Loc(1,3)    - phaseInitialCoord(2,1);
        delT03(1,5) =    tracker3Loc(1,4)    - phaseInitialCoord(2,2);
        delT03(1,6) =     delZP2(1,3)        - phaseInitialCoord(2,3);
        euclideanDistP2(1, 3) = sqrt( (delT03(1, 4).^2) + (delT03(1, 5).^2) + (delT03(1, 6).^2));
    for counter08_L = 2: frameTotPhase2
        delT03(counter08_L, 4) = tracker3Loc(counter08_L, 3) - tracker3Loc(counter08_L-1, 3);
        delT03(counter08_L, 5) = tracker3Loc(counter08_L, 4) - tracker3Loc(counter08_L-1, 4);
        delT03(counter08_L, 6) =  delZP2(counter08_L, 3)    -   delZP2(counter08_L-1, 1);
        euclideanDistP2(counter08_L, 3) = sqrt( (delT03(counter08_L, 4).^2) + (delT03(counter08_L, 5).^2) + (delT03(counter08_L, 6).^2));
    end

elseif counter08 == 4
    for counter08_M = 1: frameTotPhase1
        delXP1(counter08_M, 4) = tracker4Loc(counter08_M,1) - centersAverage(1,1);
        delYP1(counter08_M, 4) = centersAverage(1,2) - tracker4Loc(counter08_M,2);
        delZP1(counter08_M, 4) = sqrt((centersAverage(1,3).^2)-(delXP1(counter08_M, 4).^2)-(delYP1(counter08_M, 4).^2));
    end
        delT04(1,1) =    tracker4Loc(1,1)    - phaseInitialCoord(1,1);
        delT04(1,2) =    tracker4Loc(1,2)    - phaseInitialCoord(1,2);
        delT04(1,3) =     delZP1(1,4)        - phaseInitialCoord(1,3);
        euclideanDistP1(1, 4) = sqrt( (delT04(1, 1).^2) + (delT04(1, 2).^2) + (delT04(1, 3).^2));
    for counter08_N = 2: frameTotPhase1
        delT04(counter08_N, 1) = tracker4Loc(counter08_N, 1) - tracker4Loc(counter08_N-1, 1);
        delT04(counter08_N, 2) = tracker4Loc(counter08_N, 2) - tracker4Loc(counter08_N-1, 2);
        delT04(counter08_N, 3) =  delZP1(counter08_N, 4)    -   delZP1(counter08_N-1, 4);
        euclideanDistP1(counter08_N, 4) = sqrt( (delT04(counter08_N, 1).^2) + (delT04(counter08_N, 2).^2) + (delT04(counter08_N, 3).^2));
    end


    for counter08_O = 1: frameTotPhase2
        delXP2(counter08_O, 4) = tracker4Loc(counter08_O,3) - centersAverage(1,1);
        delYP2(counter08_O, 4) = centersAverage(1,2) - tracker4Loc(counter08_O,4);
        delZP2(counter08_O, 4) = sqrt((centersAverage(1,3).^2)-(delXP2(counter08_O, 4).^2)-(delYP2(counter08_O, 4).^2));
    end
        delT04(1,4) =    tracker4Loc(1,3)    - phaseInitialCoord(2,1);
        delT04(1,5) =    tracker4Loc(1,4)    - phaseInitialCoord(2,2);
        delT04(1,6) =     delZP2(1,4)        - phaseInitialCoord(2,3);
        euclideanDistP2(1, 4) = sqrt( (delT04(1, 4).^2) + (delT04(1, 5).^2) + (delT04(1, 6).^2));
    for counter08_P = 2: frameTotPhase2
        delT04(counter08_P, 4) = tracker4Loc(counter08_P, 3) - tracker4Loc(counter08_P-1, 3);
        delT04(counter08_P, 5) = tracker4Loc(counter08_P, 4) - tracker4Loc(counter08_P-1, 4);
        delT04(counter08_P, 6) =  delZP2(counter08_P, 4)    -   delZP2(counter08_P-1, 1);
        euclideanDistP2(counter08_P, 4) = sqrt( (delT04(counter08_P, 4).^2) + (delT04(counter08_P, 5).^2) + (delT04(counter08_P, 6).^2));
    end

elseif counter08 == 5
    for counter08_Q = 1: frameTotPhase1
        delXP1(counter08_Q, 5) = tracker5Loc(counter08_Q,1) - centersAverage(1,1);
        delYP1(counter08_Q, 5) = centersAverage(1,2) - tracker5Loc(counter08_Q,2);
        delZP1(counter08_Q, 5) = sqrt((centersAverage(1,3).^2)-(delXP1(counter08_Q, 5).^2)-(delYP1(counter08_Q, 5).^2));
    end
```

```matlab
        delT05(1,1) =    tracker5Loc(1,1)   - phaseInitialCoord(1,1);
        delT05(1,2) =    tracker5Loc(1,2)   - phaseInitialCoord(1,2);
        delT05(1,3) =      delZP1(1,5)      - phaseInitialCoord(1,3);
        euclideanDistP1(1, 5) = sqrt( (delT05(1, 1).^2) + (delT05(1, 2).^2) + (delT05(1, 3).^2));
      for counter08_R = 2: frameTotPhase1
        delT05(counter08_R, 1) = tracker5Loc(counter08_R, 1) - tracker5Loc(counter08_R-1, 1);
        delT05(counter08_R, 2) = tracker5Loc(counter08_R, 2) - tracker5Loc(counter08_R-1, 2);
        delT05(counter08_R, 3) =  delZP1(counter08_R, 5)   -   delZP1(counter08_R-1, 5);
        euclideanDistP1(counter08_R, 5) = sqrt( (delT05(counter08_R, 1).^2) + (delT05(counter08_R, 2).^2) + (delT05(counter08_R, 3).^2));
      end


      for counter08_S = 1: frameTotPhase2
        delXP2(counter08_S, 5) = tracker5Loc(counter08_S,3) - centersAverage(1,1);
        delYP2(counter08_S, 5) = centersAverage(1,2) - tracker5Loc(counter08_S,4);
        delZP2(counter08_S, 5) = sqrt((centersAverage(1,3).^2)-(delXP2(counter08_S, 5).^2)-(delYP2(counter08_S, 5).^2));
      end
        delT05(1,4) =    tracker5Loc(1,3)   - phaseInitialCoord(2,1);
        delT05(1,5) =    tracker5Loc(1,4)   - phaseInitialCoord(2,2);
        delT05(1,6) =      delZP2(1,5)      - phaseInitialCoord(2,3);
        euclideanDistP2(1, 5) = sqrt( (delT05(1, 4).^2) + (delT05(1, 5).^2) + (delT05(1, 6).^2));
      for counter08_T = 2: frameTotPhase2
        delT05(counter08_T, 4) = tracker5Loc(counter08_T, 3) - tracker5Loc(counter08_T-1, 3);
        delT05(counter08_T, 5) = tracker5Loc(counter08_T, 4) - tracker5Loc(counter08_T-1, 4);
        delT05(counter08_T, 6) =  delZP2(counter08_T, 5)   -   delZP2(counter08_T-1, 1);
        euclideanDistP2(counter08_T, 5) = sqrt( (delT05(counter08_T, 4).^2) + (delT05(counter08_T, 5).^2) + (delT05(counter08_T, 6).^2));
      end

elseif counter08 == 6
   for counter08_U = 1: frameTotPhase1
        delXP1(counter08_U, 6) = tracker6Loc(counter08_U,1) - centersAverage(1,1);
        delYP1(counter08_U, 6) = centersAverage(1,2) - tracker6Loc(counter08_U,2);
        delZP1(counter08_U, 6) = sqrt((centersAverage(1,3).^2)-(delXP1(counter08_U, 6).^2)-(delYP1(counter08_U, 6).^2));
      end
        delT06(1,1) =    tracker6Loc(1,1)   - phaseInitialCoord(1,1);
        delT06(1,2) =    tracker6Loc(1,2)   - phaseInitialCoord(1,2);
        delT06(1,3) =      delZP1(1,6)      - phaseInitialCoord(1,3);
        euclideanDistP1(1, 6) = sqrt( (delT06(1, 1).^2) + (delT06(1, 2).^2) + (delT06(1, 3).^2));
      for counter08_V = 2: frameTotPhase1
        delT06(counter08_V, 1) = tracker6Loc(counter08_V, 1) - tracker6Loc(counter08_V-1, 1);
        delT06(counter08_V, 2) = tracker6Loc(counter08_V, 2) - tracker6Loc(counter08_V-1, 2);
        delT06(counter08_V, 3) =  delZP1(counter08_V, 6)   -   delZP1(counter08_V-1, 6);
        euclideanDistP1(counter08_V, 6) = sqrt( (delT06(counter08_V, 1).^2) + (delT06(counter08_V, 2).^2) + (delT06(counter08_V, 3).^2));
      end


      for counter08_W = 1: frameTotPhase2
        delXP2(counter08_W, 6) = tracker6Loc(counter08_W,3) - centersAverage(1,1);
        delYP2(counter08_W, 6) = centersAverage(1,2) - tracker6Loc(counter08_W,4);
        delZP2(counter08_W, 6) = sqrt((centersAverage(1,3).^2)-(delXP2(counter08_W, 6).^2)-(delYP2(counter08_W, 6).^2));
      end
        delT06(1,4) =    tracker6Loc(1,3)   - phaseInitialCoord(2,1);
        delT06(1,5) =    tracker6Loc(1,4)   - phaseInitialCoord(2,2);
        delT06(1,6) =      delZP2(1,6)      - phaseInitialCoord(2,3);
        euclideanDistP2(1, 6) = sqrt( (delT06(1, 4).^2) + (delT06(1, 5).^2) + (delT06(1, 6).^2));
      for counter08_X = 2: frameTotPhase2
        delT06(counter08_X, 4) = tracker6Loc(counter08_X, 3) - tracker6Loc(counter08_X-1, 3);
        delT06(counter08_X, 5) = tracker6Loc(counter08_X, 4) - tracker6Loc(counter08_X-1, 4);
        delT06(counter08_X, 6) =  delZP2(counter08_X, 6)   -   delZP2(counter08_X-1, 1);
        euclideanDistP2(counter08_X, 6) = sqrt( (delT06(counter08_X, 4).^2) + (delT06(counter08_X, 5).^2) + (delT06(counter08_X, 6).^2));
      end

elseif counter08 == 7
   for counter08_Y = 1: frameTotPhase1
        delXP1(counter08_Y, 7) = tracker7Loc(counter08_Y,1) - centersAverage(1,1);
        delYP1(counter08_Y, 7) = centersAverage(1,2) - tracker7Loc(counter08_Y,2);
        delZP1(counter08_Y, 7) = sqrt((centersAverage(1,3).^2)-(delXP1(counter08_Y, 7).^2)-(delYP1(counter08_Y, 7).^2));
      end
```

```matlab
        delT07(1,1) =    tracker7Loc(1,1)   - phaseInitialCoord(1,1);
        delT07(1,2) =    tracker7Loc(1,2)   - phaseInitialCoord(1,2);
        delT07(1,3) =    delZP1(1,7)        - phaseInitialCoord(1,3);
        euclideanDistP1(1, 7) = sqrt( (delT07(1, 1).^2) + (delT07(1, 2).^2) + (delT07(1, 3).^2));
    for counter08_Z = 2: frameTotPhase1
        delT07(counter08_Z, 1) = tracker7Loc(counter08_Z, 1) - tracker7Loc(counter08_Z-1, 1);
        delT07(counter08_Z, 2) = tracker7Loc(counter08_Z, 2) - tracker7Loc(counter08_Z-1, 2);
        delT07(counter08_Z, 3) =  delZP1(counter08_Z, 7)   -   delZP1(counter08_Z-1, 7);
        euclideanDistP1(counter08_Z, 7) = sqrt( (delT07(counter08_Z, 1).^2) + (delT07(counter08_Z, 2).^2) + (delT07(counter08_Z, 3).^2));
    end


    for counter08_AA = 1: frameTotPhase2
        delXP2(counter08_AA, 7) = tracker7Loc(counter08_AA,3) - centersAverage(1,1);
        delYP2(counter08_AA, 7) = centersAverage(1,2) - tracker7Loc(counter08_AA,4);
        delZP2(counter08_AA, 7) = sqrt((centersAverage(1,3).^2)-(delXP2(counter08_AA, 7).^2)-(delYP2(counter08_AA, 7).^2));
    end
        delT07(1,4) =    tracker7Loc(1,3)   - phaseInitialCoord(2,1);
        delT07(1,5) =    tracker7Loc(1,4)   - phaseInitialCoord(2,2);
        delT07(1,6) =    delZP2(1,7)        - phaseInitialCoord(2,3);
        euclideanDistP2(1, 7) = sqrt( (delT07(1, 4).^2) + (delT07(1, 5).^2) + (delT07(1, 6).^2));
    for counter08_BB = 2: frameTotPhase2
        delT07(counter08_BB, 4) = tracker7Loc(counter08_BB, 3) - tracker7Loc(counter08_BB-1, 3);
        delT07(counter08_BB, 5) = tracker7Loc(counter08_BB, 4) - tracker7Loc(counter08_BB-1, 4);
        delT07(counter08_BB, 6) =  delZP2(counter08_BB, 7)   -   delZP2(counter08_BB-1, 1);
        euclideanDistP2(counter08_BB, 7) = sqrt( (delT07(counter08_BB, 4).^2) + (delT07(counter08_BB, 5).^2) + (delT07(counter08_BB, 6).^2));
    end

elseif counter08 == 8
    for counter08_CC = 1: frameTotPhase1
        delXP1(counter08_CC, 8) = tracker8Loc(counter08_CC,1) - centersAverage(1,1);
        delYP1(counter08_CC, 8) = centersAverage(1,2) - tracker8Loc(counter08_CC,2);
        delZP1(counter08_CC, 8) = sqrt((centersAverage(1,3).^2)-(delXP1(counter08_CC, 8).^2)-(delYP1(counter08_CC, 8).^2));
    end
        delT08(1,1) =    tracker8Loc(1,1)   - phaseInitialCoord(1,1);
        delT08(1,2) =    tracker8Loc(1,2)   - phaseInitialCoord(1,2);
        delT08(1,3) =    delZP1(1,8)        - phaseInitialCoord(1,3);
        euclideanDistP1(1, 8) = sqrt( (delT08(1, 1).^2) + (delT08(1, 2).^2) + (delT08(1, 3).^2));
    for counter08_DD = 2: frameTotPhase1
        delT08(counter08_DD, 1) = tracker8Loc(counter08_DD, 1) - tracker8Loc(counter08_DD-1, 1);
        delT08(counter08_DD, 2) = tracker8Loc(counter08_DD, 2) - tracker8Loc(counter08_DD-1, 2);
        delT08(counter08_DD, 3) =  delZP1(counter08_DD, 8)   -   delZP1(counter08_DD-1, 8);
        euclideanDistP1(counter08_DD, 8) = sqrt( (delT08(counter08_DD, 1).^2) + (delT08(counter08_DD, 2).^2) + (delT08(counter08_DD, 3).^2));
    end


    for counter08_EE = 1: frameTotPhase2
        delXP2(counter08_EE, 8) = tracker8Loc(counter08_EE,3) - centersAverage(1,1);
        delYP2(counter08_EE, 8) = centersAverage(1,2) - tracker8Loc(counter08_EE,4);
        delZP2(counter08_EE, 8) = sqrt((centersAverage(1,3).^2)-(delXP2(counter08_EE, 8).^2)-(delYP2(counter08_EE, 8).^2));
    end
        delT08(1,4) =    tracker8Loc(1,3)   - phaseInitialCoord(2,1);
        delT08(1,5) =    tracker8Loc(1,4)   - phaseInitialCoord(2,2);
        delT08(1,6) =    delZP2(1,8)        - phaseInitialCoord(2,3);
        euclideanDistP2(1, 8) = sqrt( (delT08(1, 4).^2) + (delT08(1, 5).^2) + (delT08(1, 6).^2));
    for counter08_FF = 2: frameTotPhase2
        delT08(counter08_FF, 4) = tracker8Loc(counter08_FF, 3) - tracker8Loc(counter08_FF-1, 3);
        delT08(counter08_FF, 5) = tracker8Loc(counter08_FF, 4) - tracker8Loc(counter08_FF-1, 4);
        delT08(counter08_FF, 6) =  delZP2(counter08_FF, 8)   -   delZP2(counter08_FF-1, 1);
        euclideanDistP2(counter08_FF, 8) = sqrt( (delT08(counter08_FF, 4).^2) + (delT08(counter08_FF, 5).^2) + (delT08(counter08_FF, 6).^2));
    end

elseif counter08 == 9
    for counter08_GG = 1: frameTotPhase1
        delXP1(counter08_GG, 9) = tracker9Loc(counter08_GG,1) - centersAverage(1,1);
```

```matlab
        delYP1(counter08_GG, 9) = centersAverage(1,2) - tracker9Loc(counter08_GG,2);
        delZP1(counter08_GG, 9) = sqrt((centersAverage(1,3).^2)-(delXP1(counter08_GG, 9).^2)-(delYP1(counter08_GG, 9).^2));
    end
        delT09(1,1) =    tracker9Loc(1,1)   - phaseInitialCoord(1,1);
        delT09(1,2) =    tracker9Loc(1,2)   - phaseInitialCoord(1,2);
        delT09(1,3) =     delZP1(1,9)       - phaseInitialCoord(1,3);
        euclideanDistP1(1, 9) = sqrt( (delT09(1, 1).^2) + (delT09(1, 2).^2) + (delT09(1, 3).^2));
    for counter08_HH = 2: frameTotPhase1
        delT09(counter08_HH, 1) = tracker9Loc(counter08_HH, 1) - tracker9Loc(counter08_HH-1, 1);
        delT09(counter08_HH, 2) = tracker9Loc(counter08_HH, 2) - tracker9Loc(counter08_HH-1, 2);
        delT09(counter08_HH, 3) =  delZP1(counter08_HH, 9)   -   delZP1(counter08_HH-1, 9);
        euclideanDistP1(counter08_HH, 9) = sqrt( (delT09(counter08_HH, 1).^2) + (delT09(counter08_HH, 2).^2) + (delT09(counter08_HH,
3).^2));
    end


    for counter08_II = 1: frameTotPhase2
        delXP2(counter08_II, 9) = tracker9Loc(counter08_II,3) - centersAverage(1,1);
        delYP2(counter08_II, 9) = centersAverage(1,2) - tracker9Loc(counter08_II,4);
        delZP2(counter08_II, 9) = sqrt((centersAverage(1,3).^2)-(delXP2(counter08_II, 9).^2)-(delYP2(counter08_II, 9).^2));
    end
        delT09(1,4) =    tracker9Loc(1,3)   - phaseInitialCoord(2,1);
        delT09(1,5) =    tracker9Loc(1,4)   - phaseInitialCoord(2,2);
        delT09(1,6) =     delZP2(1,9)       - phaseInitialCoord(2,3);
        euclideanDistP2(1, 9) = sqrt( (delT09(1, 4).^2) + (delT09(1, 5).^2) + (delT09(1, 6).^2));
    for counter08_JJ = 2: frameTotPhase2
        delT09(counter08_JJ, 4) = tracker9Loc(counter08_JJ, 3) - tracker9Loc(counter08_JJ-1, 3);
        delT09(counter08_JJ, 5) = tracker9Loc(counter08_JJ, 4) - tracker9Loc(counter08_JJ-1, 4);
        delT09(counter08_JJ, 6) =  delZP2(counter08_JJ, 9)   -   delZP2(counter08_JJ-1, 1);
        euclideanDistP2(counter08_JJ, 9) = sqrt( (delT09(counter08_JJ, 4).^2) + (delT09(counter08_JJ, 5).^2) + (delT09(counter08_JJ, 6).^2));
    end

  elseif counter08 == 10
    for counter08_KK = 1: frameTotPhase1
        delXP1(counter08_KK, 10) = tracker10Loc(counter08_KK,1) - centersAverage(1,1);
        delYP1(counter08_KK, 10) = centersAverage(1,2) - tracker10Loc(counter08_KK,2);
        delZP1(counter08_KK, 10) = sqrt((centersAverage(1,3).^2)-(delXP1(counter08_KK, 10).^2)-(delYP1(counter08_KK, 10).^2));
    end
        delT10(1,1) =    tracker10Loc(1,1)   - phaseInitialCoord(1,1);
        delT10(1,2) =    tracker10Loc(1,2)   - phaseInitialCoord(1,2);
        delT10(1,3) =     delZP1(1,10)       - phaseInitialCoord(1,3);
        euclideanDistP1(1, 10) = sqrt( (delT10(1, 1).^2) + (delT10(1, 2).^2) + (delT10(1, 3).^2));
    for counter08_LL = 2: frameTotPhase1
        delT10(counter08_LL, 1) = tracker10Loc(counter08_LL, 1) - tracker10Loc(counter08_LL-1, 1);
        delT10(counter08_LL, 2) = tracker10Loc(counter08_LL, 2) - tracker10Loc(counter08_LL-1, 2);
        delT10(counter08_LL, 3) =  delZP1(counter08_LL, 10)   -   delZP1(counter08_LL-1, 10);
        euclideanDistP1(counter08_LL, 10) = sqrt( (delT10(counter08_LL, 1).^2) + (delT10(counter08_LL, 2).^2) + (delT10(counter08_LL,
3).^2));
    end


    for counter08_MM = 1: frameTotPhase2
        delXP2(counter08_MM, 10) = tracker10Loc(counter08_MM,3) - centersAverage(1,1);
        delYP2(counter08_MM, 10) = centersAverage(1,2) - tracker10Loc(counter08_MM,4);
        delZP2(counter08_MM, 10) = sqrt((centersAverage(1,3).^2)-(delXP2(counter08_MM, 10).^2)-(delYP2(counter08_MM, 10).^2));
    end
        delT10(1,4) =    tracker10Loc(1,3)   - phaseInitialCoord(2,1);
        delT10(1,5) =    tracker10Loc(1,4)   - phaseInitialCoord(2,2);
        delT10(1,6) =     delZP2(1,10)       - phaseInitialCoord(2,3);
        euclideanDistP2(1, 10) = sqrt( (delT10(1, 4).^2) + (delT10(1, 5).^2) + (delT10(1, 6).^2));
    for counter08_NN = 2: frameTotPhase2
        delT10(counter08_NN, 4) = tracker10Loc(counter08_NN, 3) - tracker10Loc(counter08_NN-1, 3);
        delT10(counter08_NN, 5) = tracker10Loc(counter08_NN, 4) - tracker10Loc(counter08_NN-1, 4);
        delT10(counter08_NN, 6) =  delZP2(counter08_NN, 10)   -   delZP2(counter08_NN-1, 1);
        euclideanDistP2(counter08_NN, 10) = sqrt( (delT10(counter08_NN, 4).^2) + (delT10(counter08_NN, 5).^2) + (delT10(counter08_NN,
6).^2));
    end
```

```matlab
        end %end elseif loop that performs a calculation based on the current tracker

    end %end the for loop which repeats the elseif loop until calculations have been performed on all trackers


%Next step: sum up all previous rows for each tracker column

%remember euclideanSumP1 = zeros(frameTotPhase1,prompt_06_AMat);
%remember euclideanSumP2 = zeros(frameTotPhase2,prompt_06_BMat);

%sum up all previous rows (frames) for each tracker before the current frame. (Phase 1 only) --> original code in williams translation of POI
incorrectly left out the first row as zeros (which had to be fixed by hand in post processing. When comparing to original, note that williams
matrix had fliped rows and columns compared to the definition of row and column in this code. ie. his rows represented trackers and columns
frames. My code has rows as Frames and trackers as columns for consistancy with previous matrices. (His code was inconsitant)
for counter08_FinalA = 1:prompt_06_AMat %prompt a is the number of trackers in phase 1
    euclideanSumP1(1, counter08_FinalA)= euclideanDistP1(1, counter08_FinalA); %the first row of euclidean sum (representing distance from
each trackers first frame to the phase center) is identical to the euclidean distance matrix

    for counter08_FinalB = 2:frameTotPhase1 %counter B rotates through frames for the tracker specified by counterA. start with the row for
frame 2 because frame 1 is recorded by the previous line
        euclideanSumP1(counter08_FinalB, counter08_FinalA) = euclideanSumP1(counter08_FinalB - 1, counter08_FinalA) +
euclideanDistP1(counter08_FinalB, counter08_FinalA); %every value starting at the second row onward is the sum of the current and previous
rows of euclidean distance. therefore,to save time, the previous row of euclidean sum (which is in itself a sum of previous rows) simply has to be
added to the current row of euclidean distance to obtain the current row of euclidean sum.
    end                                          %note the -1 is in the columns for this code , but in the row for williams.
end

%Repeat the above calculation for phase 2
for counter08_FinalC = 1:prompt_06_BMat %prompt b is the number of trackers in phase 2
    euclideanSumP2(1, counter08_FinalC)= euclideanDistP2(1, counter08_FinalC);

    for counter08_FinalD = 2: frameTotPhase2
        euclideanSumP2(counter08_FinalD, counter08_FinalC) = euclideanSumP2(counter08_FinalD -1, counter08_FinalC) +
euclideanDistP2(counter08_FinalD, counter08_FinalC);
    end
end



%% Calculation Results
    fprintf('x,y,z coordinates for the initiation point of phase 1 (row 1) and phase 2 (row 2) \n');
    disp(phaseInitialCoord); %phase center coordinates in 3-D space for rows phase 1 and phase 2, columns x,y,z

    fprintf('Phase 1 matrix with one column for each tracker, and rows representing distance moved since previous frame \n');
    fprintf('the first row is Euclidean Distance between the first frame and phase initiation coordinate, all other rows are distance from previous
frame \n');
    fprintf('subsequent rows compare the distance between the tracker coordinates on a given frame and its coordinates in the previous frame \n');
    disp(euclideanDistP1)

    fprintf('Phase 2 Euclidean Distance matrix with a column for each tracker \n');
    disp(euclideanDistP2)

    fprintf('Phase 1 Total Distance matrix to be plotted\n');
    fprintf('each row represents the total distance from the phase center that the tracker (column) has moved by that frame \n')
    disp(euclideanSumP1)

    fprintf('Phase 2 Total Distance matrix to be plotted\n');
    disp(euclideanSumP2)

%% Plots

%create matrix with values from start of phase 1 to phase 2
    %counter09 = 1;
    counter10 = 1;
    counter11 = phase1Start; %counter 11 will be incremented starting at the frame number where phase 1 was first seen
    %counter12 = 1;
    counter13 = 1;
    counter14 = phase2Start; %counter 11 will be incremented starting at the frame number where phase 2 was first seen
```

```matlab
    %counter15
    %counter16

%initialize variables
    FrameRefPhase1 = zeros(1000, 2); %initialize a really big matrix to represent x,y coordinate pairs (frame number of the phase 1 plot)
    FrameRefPhase2 = zeros(1000, 2); %initialize a really big matrix to represent x,y coordinate pairs (frame number of the phase 2 plot)


    %now fill the above matrices with x,y pairs taken from data in the euclidean sum matrices from the calculations step, x coordinates must be
generated using a counter to scroll through known frames in a phase nd record frame number


%phase1 x-axis coordinates (frame#) are saved to each row in the first column of frame reference phase 1
for counter09 = phase1Start : phase1End  %for each frame in the phase eg. frame 99 --> frame 1000 counter08 will count up
    FrameRefPhase1(counter10, 1) = counter11; %counter09 assumes the start frame is 1 and counts up, saving the value of counter10 to a new
colmn each loop. Counter10's starting value is the phase1 start, and it also counts up with each loop until reaching phase 1 end
    counter10 = counter10 + 1; %add 1 to the current row to move to the next row  (which starts at the start frame number)
    counter11 = counter11 + 1; %add 1 to the current frame , this value is to be recorded in the next row
end  % result FrameRefPhase1 = (phase1Start, phase1Start+1, phase1Start+2,..... phase1End) can be matched with point translations from center
to create an x,y graph.

%phase2 x-axis coordinates (frame#) are saved to each row in the first column of frame reference phase 2
for counter12 = phase2Start : phase2End %same as above
    FrameRefPhase2(counter13, 1) = counter14;
    counter13 = counter13 + 1;
    counter14 = counter14 + 1;
end

%phase1 y-axis coordinates (average distance traveled since phase 1 center - measured in pixels)
for counter15 = 1 : frameTotPhase1 %cycle through frames (rows) and fill in the second column of each row with a y coordinate
    FrameRefPhase1(counter15, 2) = mean(euclideanSumP1(counter15, :), 2); %fill in each row of the second column of the frame ref phase 1
matrix with the average of all trackers for the current row in the euclidean sum matrix. Therefore the distances of all trackers saved to any given
frame are averaged together and saved as a single value (2 indicates averaging rows not columns)
end

%phase2 y-axis coordinates (average distance traveled since phase 2 center - measured in pixels)
for counter16 = 1 : frameTotPhase2 %cycle through and fill in each row
    FrameRefPhase2(counter16, 2) = mean(euclideanSumP2(counter16, :), 2); % same as phase 1 y axis comment
end


    %now plot the x,y coordinate pairs on a graph in a new figure window


%Phase 1 Plot
for counter17 = 1:frameTotPhase1
    plot(FrameRefPhase1(counter17, 1), FrameRefPhase1(counter17, 2), 'o', 'color', rand(1,3)); %plot the x,y pair for each row as a point
    hold on
end
linearRegP1 = fitlm(FrameRefPhase1(1:frameTotPhase1, 1), FrameRefPhase1(1:frameTotPhase1, 2)); % Linear Regression to obtain the point
where the trend line crosses the x axis. Should be within 1 or two frames of phase 1 start. the x intercept indicates the true start of phase 1 and is
used to obtain delay time when compared to the same point in phase 2
plot(linearRegP1,'color', rand(1,3));

    %experimental phase 1 plot without error bars
    %x = FrameRefPhase1(1:frameTotPhase1, 1);
    %y = FrameRefPhase1(1:frameTotPhase1, 2);
    %X = [ones(length(x),1) FrameRefPhase1(:, 1)];
    %b = X\y;
    %yCalc2 = X*b;
    %hold on
    %plot(x,yCalc2)

%Phase 2 Plot
for counter18 = 1:frameTotPhase2
    plot(FrameRefPhase2(counter18, 1), FrameRefPhase2(counter18, 2), 'o', 'color', rand(1,3)); %plot the x,y pair for each row as a point
    hold on
end
linearRegP2 = fitlm(FrameRefPhase2(1:frameTotPhase2, 1), FrameRefPhase2(1:frameTotPhase2, 2)); % Linear Regression
plot(linearRegP2,'color', rand(1,3));
```

hold off

%Calculate and display Trend Line Equation for phases 1 and 2

```
% Use Linear Regression 1 to calculate trend line slope and y intercept
m1 = linearRegP1.Coefficients.Estimate(2);
b1 = linearRegP1.Coefficients.Estimate(1);
fprintf('y1 = %.3f x1 + %.3f \n', m1, b1);

% Use Linear Regression 2 to calculate trend line slope and y intercept
m2 = linearRegP2.Coefficients.Estimate(2);
b2 = linearRegP2.Coefficients.Estimate(1);
fprintf('y2 = %.3f x2 + %.3f \n', m2, b2);

% Enter the Frame Rate Found in the cine viewer editior
prompt_FPS = 'input the frame rate '; %creates the label for the dialogue box
frameRate = inputdlg(prompt_FPS); %save users response (file name)
frameRateMat = cell2mat(frameRate);

% Calculate delay time by solving for x intercept (the approximate fraction of a frame when recalescence began) for phase 1 and 2
% (0=mx+b) then subtracting phase 1 from phase 2 x intercepts  and dividing by the fps of the video to get time
syms t1 t2
delayTime = vpa((vpasolve(m2 * t2 + b2 == 0, t2) - vpasolve(m1 * t1 + b1 == 0, t1))/frameRateMat, 4);
fprintf('delay time between two phases is %f \n\n\n', delayTime);


%Check for correctness against actual phases
checkP1 = -1.*b1./m1 ;
fprintf('Estimated Phase 1 start is %f \n', checkP1);
fprintf('Actual Phase 1 start is %f \n', phase1Start);

checkP2 = -1.*b2./m2 ;
fprintf('Estimated Phase 2 start is %f \n', checkP2);
fprintf('Actual Phase 2 start is %f \n', phase2Start);
```

%%% Plot results

```
    fprintf('Congrats on a successful run! \n')
    fprintf('You have now plotted average tracker translation distance by frame \n\n');
    fprintf('x,y coordinates for phase 1 \n')
        disp(FrameRefPhase1((1:frameTotPhase1), :))
    fprintf('x,y coordinates for phase 2 \n')
        disp(FrameRefPhase2((1:frameTotPhase2), :))
```

%%% Don't like results? Trim data and replot? (Experimental)

```
%duplicate plotted matrices
%FrameRefPhase1Temp = FrameRefPhase1;
%FrameRefPhase2Temp = FrameRefPhase2;

%outer while prompt == 'Y'
%prompt - would you like to make a change?
%prompt - continue to change or restart change
    %if continue to change
      %keep frameref from previous run
    %else if restart change
      %FrameRefPhase1Temp = FrameRefPhase1;
      %FrameRefPhase2Temp = FrameRefPhase2;
```

```matlab
%inner loop to change data
    %prompt user which phase to modify
        %if phase 1
            %prompt which point would you like to modify
            %prompt which tracker would you like to remove
            %recalculate avarages (without that point)
            %FrameRefPhase1Temp = ....

        %if phase 2
            %prompt which point would you like to modify
            %prompt which tracker would you like to remove
            %recalculate averages (without that point)
            %FrameRefPhase1Temp = ....

        %plot graph


% save results when while loop finished and print updated plot


%% Print Results and save to a file

%create a new folder to store data from this run
    prompt_97_SaveFolder = 'Input the name of the "Folder" you want to save: ';  %creates the label for the dialogue box that asks what name to
save the the plot as
    newFolder = inputdlg(prompt_97_SaveFolder);
    dirName = sprintf('%s', newFolder{1,1}); %automatically specifies the directory for the the new folder (%s) to the desktop
    mkdir('C:\Users\Name\Desktop', dirName);

%move all photos from the current run to the new folder
    newDir = sprintf('C:\\Users\\Name\\Desktop\\%s', dirName); %double slashes indicate directory when used with sprint f
    cd C:\Users\Name\Desktop\Matlab\Photos %open the Photos folder where photos from the current run are temporarily stored
    movefile('*.jpg', newDir ); %move all current photos to the folder for the current run on the desktop (with the newDir directory)

%Save plot with phases 1 and 2 to desktop
    cd(newDir);
    prompt_98_SavePlot = 'Input the name of the "Plot" you want to save: ';  %creates the label for the dialogue box that asks what name to save
the the plot as
    savePlot = inputdlg(prompt_98_SavePlot); %opens a dialogue box with the label specified above that allows the user to enter a "name"
    savePlotMat = sprintf('%s.jpg',savePlot{1,1}); % "sprintf" is a function that prints a sentance from a set of variables using %s (letter) ot %d
(number) calls. In this case the %s calls in the word saved to first variable after the comma eg.the "name" entered in the save plot matrix.
therefore "%s .jpg"  = "name .jpg"
    exportgraphics(gcf, savePlotMat) %get the current figure (gcf), and save that figure (plot) to desktop with a file name identified by
savePlotMat in the line above.

%ask user what file name the backup file should be saved as
    cd(newDir);
    prompt_99_SaveFile = 'Input the name of the "ISS_EML" file you want to save: ';  %creates the label for the dialogue box
    saveName = inputdlg(prompt_99_SaveFile); %saves your file name as a variable
    saveNameMat = sprintf('%s.mat',saveName{1,1}); %converts your file name to a string of text (%s) and adds the .mat ending to ensure it is
saved as a mat file
    save(saveNameMat) %backs up all variables to a new file on the desktop with the string name you created.

%return to the original directory
    cd 'C:\Users\Name\Desktop'; %change directory to users desktop (change 'user' based on computer being used)

fprintf('Thank you for running Delay_Times.mat good luck on the data processing! \n')
```